CITY

... provides an accessible yet critical introduction to one of the key areas in human geography. Always at the heart of discussions in social theory, the definition and specification of the city nonetheless remains elusive. In this volume, Phil Hubbard locates the concept of the city within current traditions of social thought, providing a basis for understanding its varying usages and meanings through a critical discussion of the contribution of major theories and thinkers. This book thus offers a distinctive and timely intervention in debates in urban theory by suggesting new ways that students and scholars in sociology, geography, urban studies, planning and politics can make sense of the city.

Written in a lively and accessible style, the individual chapters of City offer a thematic overview of some important ways of approaching cities, whether as imagined realms, lived-in places, networks of association or spaces of flow. Situating these traditions within the rich heritage of urban studies and urban sociology, the book develops the argument that none of these approaches, when taken alone, helps us grasp the specificity of the urban, but that each is vital for grasping the materiality of cities. The book thus spells out the case for a renewed urban geography, suggesting that it is only by combining these different ways of approaching the city that we can begin to understand the relational materiality of urban life.

Phil Hubbard is Reader in Urban Social Geography at Loughborough University.

Key Ideas in Geography

SERIES EDITORS: SARAH HOLLOWAY, LOUGHBOROUGH UNIVERSITY AND GILL VALENTINE, LEEDS UNIVERSITY

The *Key Ideas in Geography* series will provide strong, original and accessible texts on important spatial concepts for academics and students working in the fields of geography, sociology and anthropology, as well as the interdisciplinary fields of urban and rural studies, development and cultural studies. Each text will locate a key idea within its traditions of thought, provide grounds for understanding its various usages and meanings, and offer critical discussion of the contribution of relevant authors and thinkers.

Published

Nature
NOEL CASTREE

City
PHIL HUBBARD

Home
ALISON BLUNT AND ROBYN DOWLING

Forthcoming

Scale
ANDREW HEROD

Landscape
JOHN WYLIE

Environment
GEORGINA ENDFIELD

CITY

Phil Hubbard

Routledge
Taylor & Francis Group

LONDON AND NEW YORK

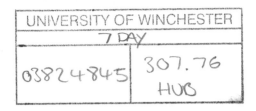
First published 2006
by Routledge
2 Park Square, Milton Park, Abingdon, Oxon OX14 4RN

Simultaneously published in the USA and Canada
by Routledge
711 Third Avenue, New York, NY 10017 (8th Floor)

Routledge is an imprint of the Taylor & Francis Group, an informa business

© 2006 Phil Hubbard

Typeset in Joanna and Scala Sans by
Keystroke, Jacaranda Lodge, Wolverhampton

British Library Cataloguing in Publication Data
A catalogue record for this book is available from the British Library

Library of Congress Cataloging in Publication Data
Hubbard, Phil.
City / Phil Hubbard.
     p. cm. — (Key ideas in geography)
Includes bibliographical references.
ISBN 0–415–33099–8 (hardcover : alk. paper) — ISBN 0–415–33100–5
(softcover : alk. paper) 1. Cities and towns. 2. Sociology, Urban.
3. Urbanization. 4. Metropolitan areas. I. Title. II. Series.

HT151.H82 2006
307.76—dc22
2006007552

ISBN10: 0–415–33099–8 (hbk)
ISBN10: 0–415–33100–5 (pbk)

ISBN13: 978–0–415–33099–2 (hbk)
ISBN13: 978–0–415–33100–5 (pbk)

For Brian and Joy

CONTENTS

LIST OF ILLUSTRATIONS AND TABLES ix
LIST OF BOXES xi
ACKNOWLEDGEMENTS xiii

Introduction 1

1 Urban theory, modern and postmodern 9

2 The represented city 59

3 The everyday city 95

4 The hybrid city 129

5 The intransitive city 164

6 The creative city 206

Conclusion 247

BIBLIOGRAPHY 249
INDEX 279

ILLUSTRATIONS AND TABLES

PLATES

1.1	*A Spring Morning, Haverstock Hill* (George Clausen, 1881)	13
1.2	Postmodern urban forms	52
2.1	Star City multiplex, Birmingham	64
2.2	The London Underground map	77
2.3	*Heart of Empire* (Neils M. Lund, 1904)	82
2.4	Poster used in promoting Leicester	90
3.1	The Seattle Sky Needle	99
3.2	Regulating space	105
3.3	Le parkour	109
3.4	Stencil art in London	110
3.5	Post-war planning	115
4.1	*Scientific American* April 1881 celebrating Edison's achievement of providing electrically powered street lighting in New York	134
4.2	An urban actor? Carpet in the Parisian gutter	148
4.3	CCTV has become a routine part of urban life	149
4.4	The return of the repressed?	155
6.1	White Cube gallery, Hoxton Square, London	220
6.2	Art, alcohol and the social spaces of London	222
6.3	Brick Lane, Spitalfields	243

FIGURES

1.1   The postmodern urban landscape   55
5.1   Growth in global airspace movements   167
5.2   Global network connectivity of advanced producer
      service firms   182
5.3   Interconnectivity of the world's twenty-five most
      important air hubs   184
5.4   Urban assets and competitive outcomes   200
6.1   Art networks: Damien Hirst located in networks of
      creativity   227

TABLES

6.1   Large US cities' creativity rankings   213
6.2   Advantages of face-to-face communication   231

BOXES

1.1	The Chicago School	25
1.2	Marxist urbanism	35
1.3	The postmodern city	50
2.1	The cinematic city	62
2.2	Urban semiotics	71
2.3	City branding	86
3.1	The flâneur	102
3.2	Situationism	108
3.3	Non-representational theory	120
4.1	Virtual cities	138
4.2	Actor Network Theory	145
4.3	Cyborg cities	156
5.1	The new mobilities paradigm	171
5.2	The informational city	178
5.3	Polycentric urban regions	197
6.1	Gay villages	215
6.2	Buzz cities	232
6.3	Spaces of diaspora	241

Acknowledgements

Given this book provides an overview of a massive literature, it is necessary to begin by acknowledging the enormous number of authors whose work could have been cited here, yet for a variety of reasons – including my own sheer ignorance – I have not referenced. The fact they have not been referenced does not mean they are insignificant to the debates I review, nor that they have not informed the writing of this book, consciously or otherwise. As such, I can only apologise to the many whose work I should have cited, as well as those who will object to my interpretation of their work. This inability to do justice to the rich diversity of scholarship in urban geography is, alas, inescapable in a book of this type where space is finite. Nonetheless, I hope I have given full credit for ideas where credit is due – something not always easy where one is surrounded by colleagues whose work inspires and enthuses. In this context, I would wish to note that certain sections of this book are related to or have emerged from works I have written with Keith Lilley, Marcus Doel, Jon Beaverstock, Peter Taylor, John Short, Rob Kitchin, Lewis Holloway, Brendan Bartley, Duncan Fuller, Tim Brown, Morag Bell and Lucy Faire (among others). Particular thanks is also due to my postgraduate students at Loughborough – Lucy Budd, Clare Blake, Julia Grosspietsch, Ulrike Waellisch – whose research has helped inform some of the material in this book. I would also wish to thank the other colleagues who have, through their own research and writing, helped inspire my work over the last few years: it is invidious to pick out individuals, but Sarah Holloway deserves to be mentioned, given she asked me to write this book in the first place and was so patient in waiting for it to arrive! Andrew and the team at Taylor and Francis have also been patient and professional throughout while Christine Firth provided the wonderfully thorough copy-editing the text desperately needed. Finally, Cath, Lucy and Oliver have put up with my constant moans, and their support and love matter more than anything else.

Plate 1.1 with kind permission of Richard Burns, Bury Art Gallery and Museum.

Figure 1.1 is redrawn with kind permission of Michael Dear and Stephen Flusty.

Plate 2.2 is reproduced with kind permission of the London Transport Museum.

Plate 2.3 is reproduced with kind permission of London Guildhall Art Gallery Collection.

Plate 2.4 is reproduced with kind permission of Leicestershire Promotions Limited.

Plate 3.1 is a work of the United States Federal Government and is reproduced under the terms of Title 17, Chapter 1, Section 105 of the US Code.

Plate 3.3 is reproduced with kind permission of Pauli Peura.

Plate 3.4 is reproduced with kind permission of Banksy.

Plate 3.5 is reproduced with kind permission of Coventry and Warwickshire Libraries Collection.

Plate 4.1 is reproduced with kind permission of Scientific American.

Plate 4.4 is reproduced with kind permission of Tim Edensor.

Figure 5.2 is reproduced with kind permission of Peter Taylor.

Figure 5.3 is reproduced with kind permission of Ben Derudder and Frank Witlox.

Figure 5.4 is redrawn with kind permission of Iain Deas and Benito Giordano.

Plate 6.2 is reproduced with kind permission of Martin Rowson.

Figure 6.1 is redrawn with kind permission of Mike Grenfell.

Plate 6.3 is reproduced with kind permission of Robert House.

All other photos are copyright the author.

INTRODUCTION

Despite its title, this is not a book about cities. Rather, it is a book about urban theory, exploring the way that sociologists, planners, architects, economists, urbanists and (particularly) geographers have sought to make sense of the urban condition. As such, the primary aim of this book is to locate the concept of 'the city' within traditions of geographic and social thought, providing a basis for understanding its varying usages and meanings. This is by no means a straightforward task, as the city is many things: a spatial location, a political entity, an administrative unit, a place of work and play, a collection of dreams and nightmares, a mesh of social relations, an agglomeration of economic activity, and so forth. To isolate some characteristic of the city that might distinguish it from its nominal counterpart, the rural, is extremely difficult given this multiplicity of meaning.

To illustrate this, we might briefly consider some definitions of what differentiates urban and rural. For some, what distinguishes cities from the countryside is their size and population density. For example, R. Davis (1973: 1) describes cities as 'concentrations of many people located close together for residential and productive purposes' while Saunders (1986: 7), in addressing this question, points out 'cities are places where large numbers of people live and work'. Yet definitions based on size are problematic given that settlements of the same size may be designated as small towns in some nations, but large villages in others. And it is certainly the case that some settlements *feel* like cities, while other – more populous – settlements do not. Hence, some propose that questions of population density or heterogeneity may be more crucial in definitions of the city. Others argue that it is the ways of life (or *cultures*) characteristic of cities that distinguish them from rural spaces, or that it is the way nature is excluded from cities that marks them off as being different from the countryside (for an overview of such debates, see Pile 1999).

It is thus possible to make distinctions between the urban and the rural on a number of different criteria (and one should not forget that the country and the city are also morphologically different, in the sense they are characterised by different built forms and layouts). However, there are many commentators who suggest the distinction between rural and urban spaces is becoming irrelevant – or at least less relevant – to the extent that concepts such as 'the urban' and 'the rural' are no longer useful for making sense of societies characterised by high levels of geographic and social mobility. Successive transport and communications innovation have, they would argue, loosened the ties that bind people to place, to the extent that it no longer makes sense to talk of rural or urban dwellers (indeed, many people cannot even be tied down as citizens of one nation, exhibiting transnational lifestyles and affinities). A related tendency is therefore to emphasise the footloose nature of contemporary life, and to emphasise the stretching of relations of all kinds across both time and space.

An unfortunate by-product of this mode of thinking is that much contemporary writing reduces the city (and, likewise, the countryside) to the status of a container or backdrop for human activities, downplaying (and, at worst, ignoring) its profligate role in shaping economic and social relations. This failure to take the city seriously means that much contemporary urban commentary actually says very little about what Soja (2000) terms the *generative* aspects of urban life. For example, much contemporary writing on cities aims to speak to the problems experienced by those living in cities, identifying ways in which urban processes can be modified, disturbed or corrected to reduce the high levels of crime, disease, fear and poverty that are characteristically associated with cities. However, whether these are problems of the city, or merely social problems that happen to be located in cities is rarely addressed. For instance, at the time of writing, riots rage in the *banlieu* of Paris and other major French cities, yet the academic commentators routinely assembled to pass judgement on these disturbances have tended to fixate on questions of cultural integration, social stigmatisation and racism. In contrast, little, if anything, has been said about the spatial specificity of the urban spaces where these riots are unfolding – notable exceptions being those newspaper articles where spurious correlations have been made between the unrest and the forms of public housing characteristic of the *banlieu*. While the look and feel of urban spaces do have important effects on behaviour, such design-centred theories offer an impoverished take on the distinctive sociality of cities (Amin et al. 2000): what is needed is urban scholarship that takes the city seriously as an object of study without lapsing into environmental determinism. Without such explorations, it is less than clear as to how we might suggest ways in which the trajectory of urban life might be changed through new ways of living, occupying or imagining cities.

The deficiencies of contemporary urban theory are sharply etched in other ways. For instance, there is an emerging literature on the creativity that is associated with cities, exploring why particular cities are at the 'cutting edge' of cultural or economic innovation. The association between particular cities and specific cultures of artistic innovation is widely noted: for example, cubism and impressionism can be traced back to *fin-de-siècle* Paris, jazz to New Orleans in the 1920s and beat poetry to San Francisco in the 1950s (P. Hall 1998). Likewise, certain cities are acknowledged to have acted as centres of economic, industrial or technical innovation (e.g. Manchester was dubbed Cottonopolis because of its centrality in nineteenth-century cotton and textiles industry, while Seattle is world renowned as a centre of high-tech and software design). Common sense thus dictates that cities foster creativity and vitality, and that cutting-edge art, fashion, music, cooking, technology, knowledge, politics and ideas tend to emanate from urban, not rural, locales. But why should this be the case? Is it simply that large cities are more likely than rural areas to be home to creative individuals? Is it that creative individuals are drawn towards big cities? Is there something about particular cities that fosters creativity? As we shall discover in Chapter 6, there is perhaps something in all of these explanations, yet all too often the materiality of the city is overlooked in accounts that emphasise individual genius, collective endeavour or historical happenstance. In essence, the urbanity of urban life is effaced: cities are written of as spaces where innovation happens, for sure, but the city becomes backdrop rather than an active participant in the making of new cultures and economies. Again, to suggest the city plays an active role in innovation is not to imply it has a deterministic influence on the trajectory of economy or society, but to argue we need to take the city more seriously if we are to articulate the importance of space in social, economic and political life.

The fact that urban theory currently appears unable to provide a useful framework for making sense of cities has been widely noted across a variety of disciplines. For example, Manuel Castells (2000) has publicly mourned the passing of urban sociology, suggesting that the theoretical imagination and vibrancy that once characterised urban studies in the 1960s and 1970s have been lost. In his opinion, there is a need for sociologists to retool with new concepts and ideas, before embarking on the hard work required to understand the changing relationship between the city and society. Simply put, he feels that contemporary sociology is content to say plenty of things about what happens in cities, yet has said little about the specificity of urban process. Mirroring this, May and Perry (2005) suggest this 'crisis' in urban sociology reflects the fragmentation of urban sociology, an

> inward collapse and retreat into a series of separate studies that draw on urban sociology but frequently without explicit credit to the discipline . . . for example, in the areas of housing, education, policy and cultural studies, gender and sexuality, crime and ethnicity.
>
> (May and Perry 2005: 343)

Writing as a geographer, Nigel Thrift (1993) likewise proclaimed an urban impasse, manifest in the loss of the urban as both a subject and object of study. In his summation, while geographers have developed a varied and rich lexicon for describing urban phenomena, this cannot overshadow the fact urban geography had, by the 1990s, been 'treading water' in theoretical terms, recycling old ideas for over two decades. Thrift's impasse is arguably still evident in the twenty-first century, and it is no doubt significant that many urban geographers now prefer to identify themselves as cultural geographers, feminist geographers, population geographers, economic geographers, social scientists and so on. Although urban geography remains one of the largest specialty research groups in both the Institute of British Geographers and the Association of American Geographers, it is also one of the most diffuse. As a result, the intellectual developments that have swept over and transformed the discipline since the mid-1990s are usually associated with cultural, economic or feminist geography, even if many of them have been urban in text and context. Contemporary geographic writing on 'fashionable' topics such as spaces of consumption, leisure, music, technology and travel is rarely associated with urban geography, despite the fact that the city is central to so many of them. Sexual geographies, for example, are not thought of as situated centrally within urban geography. Hence, as Johnston (2000: 877) suggests, 'many of the concerns formerly encapsulated within urban geography are now studied under different banners'.

It is thus possible to note a worrying trend for urban studies in general, and urban geography in particular, to be sidelined in intellectual and theoretical discussions about the relation of society and space. Simply put, urban geography is not seen to be at the forefront of innovative theoretical work in the discipline any more (Lees 2001; P. Jackson 2005). Yet at the same time, it is possible to discern a popular resurgence of interest in urbanity and urbanism. For example, there is much contemporary media discussion about the major reinvestment in the heart of cities long regarded as 'no-go' areas. Throughout the urban West, under the aegis of urban policies promoting urban 'renaissance', former manufacturing districts are being repackaged and resold; derelict waterfronts have become a locus for gentrified living and working; mega-malls, multiplexes and mega-casinos seek to capture the dollars of the new urban elite. Western cities, it seems, have gone from strength to strength: hence, there is talk of a 'new' urbanism (Katz 1994; McCann 1995), the new city (Sorkin 1992) and the new urban frontier (N. Smith 1996). The young, trendy and wealthy again clamour to invest in urban property hotspots, reversing a trend of depopulation and out-migration that had persisted for decades. Cities are deemed to be hip and happening again: some may be 'hot', others 'cool', but there is little doubt that cities are currently the 'place to be'. Of course, city living comes replete with dangers as well as pleasures, and this is perhaps part of the appeal of cities for a new generation of gentrifiers and trend-setting urbanites. Indeed, such groups often emphasise the authentic and spontaneous nature of city living, suggesting this is infinitely preferable to the predictable nature of country life. It is surely no coincidence that many of the most popular computer games of our times are urban (such as *Grand Theft Auto: Vice City*, *Warriors*, *Driver* and *Sim City*'s suburban sequel, *The Sims*).

Within the popular and policy-oriented literature on cities, there are undoubtedly many clues to be gleaned about the nature of contemporary urban life. However, rather than providing an overview of these voluminous literatures, in this book I seek to provide a more parsimonious summary of key ideas in urban theory, based chiefly on the work of academic writers in geography, sociology, planning and cultural studies. In broad terms, urban theory constitutes a series of ideas (sometimes presented as laws) about what cities are, what they do and how they work. Commonly, such ideas exist at a high level of abstraction, so that they do not pertain to individual towns or cities, but offer a more general explanation of the role that cities play in shaping socio-spatial process. Nonetheless, such theories typically emerge from particular cities at particular times, to the extent that certain cities become exemplary of particular types of urban theory, and become regarded as laboratories in which theories can be refined or 'tested'. By way of example, some of the most influential urban theories of the twentieth century sought to make generalisations about contemporary urban processes by referring to Chicago, which became both model and mould for theories of urbanism in the 'modern' industrial era. Likewise, much postmodern urban theory took shape in the Los Angeles of the late twentieth century, a city whose landscapes were taken as symptomatic of not only North American urbanisation, but also the wider restructuring of social-spatial relations in postmodern times (see Chapter 2).

Obviously, even in a book focusing on the urban as a concept or idea, there is much ground that could be covered. As such, I do not pretend that the following chapters do anything but offer a very partial overview of the huge body of work that constitutes 'urban theory'. Indeed, I deliberately skim over much of the literature on cities to focus on the ideas which I regard as most useful and making sense of the city. Perhaps inevitably, I tend to focus upon ideas which have emerged in the relatively short time I have been lecturing in higher education (since 1994), including writings by many who might broadly be described as my contemporaries. While I make no apologies for highlighting the distinctive and important contribution of a relatively select band of Anglo-American geographers, it is important that readers are aware of the inevitably partial nature of this text. Like the other books in the series, it cannot be read as the definitive guide to a particular concept; rather, it is a selective, simplified and Anglo-centric review. Notably, it rarely addresses non-Western literatures or theories developed beyond the metropolitan centres of the global north. While this occlusion is problematic – the disjuncture between Western and non-Western urban studies being a major impediment to the development of a progressive urban theory – my belief is that this overview nonetheless speaks to a general debate about cityness, albeit that this debate needs to be played out differently in particular contexts.

Consequently, I hope that – in spite of its silences and blind spots – this book serves as an informative and interesting overview of the way geographers have thought about the urban. Moreover, I hope that readers will come away appreciating why the city remains such an important concept in the production of geographical theories and knowledges, one that is pivotal in progressing our understanding of the relationship of society and space. Although my own work varies considerably in its approach and empirical specificity, this is an argument I believe in passionately, and I have often been dismayed by the inability of geographers to take the city seriously, both as an object of study and as a theoretical concept. While some might argue that we live in an ubiquitously urban world, and hence that it is time to 'do away with the urban' (cf. Hoggart 1991), this book is unequivocally an argument for the *reassertion* of the urban. In essence, this book seeks to demonstrate that the 'city' – though a disputed and often chaotic concept – is as central and important to geographical inquiry as concepts such as 'place', 'space', 'region', 'nature' or 'landscape'. In part, it also suggests that the concept of urbanity can be understood only with reference to the concept of 'rurality'.

In this book, I thus seek to offer a distinctive and timely intervention in debates in human geography by exploring the ways that scholars in sociology, geography, urban studies, planning and politics might make better sense of 'the city' by taking its spatiality seriously. Though written from a geographical perspective (and with geography students in mind), I hope that it speaks to an interdisciplinary audience. Throughout, some important concepts associated with urban theory have been highlighted in *italics*: these concepts are explained in boxes, each of which is intended as a free-standing elucidation of an important idea which may be familiar to some readers, but less so to others. Each of these boxes also includes recommendations for further reading, intended to steer readers towards fuller and more detailed explications of the material summarised in these boxes. Additionally, each chapter ends with some recommendations for those wishing to read more widely around the themes and ideas considered. As far as is possible, readers are guided towards widely available and accessible English-language texts (which, in some cases, means works that are extracted or translated in part). Together, these features are intended to help the reader navigate the text, as well as encourage a broader engagement with work in urban theory.

Though some readers will discern a chronological element, this volume is organised thematically: the intention is not to present a 'history' of urban geography or urban theory. Accordingly, the book unfolds by considering distinctive approaches to understanding the city, placing considerable emphasis on those ideas which I regard as particularly useful for making sense of the city. The book begins in Chapter 1 by delineating some key traditions in urban theory, identifying many of the key ideas that have emerged to explain the role of cities in modern and postmodern times. Detailing theories that stress the economic, social, political or cultural dimensions of city life (as well as some that consider all of these), the chapter concludes by identifying the apparent limitations of such theories in developing a truly urban theory. Chapters 2 to 5 then overview some alternative ways of thinking about cities and theorising their distinctive spatiality. In turn, these deal with ideas concerning the representation of space; the negotiation of the everyday; the blurring of nature and culture and the stretching of social relations. As will be seen, many of the ideas considered in these chapters have not explicitly been developed with the goal of progressing urban theory. Nor can any of them be said to offer a 'grand theory' of how the city works. Hence, in Chapter 6, I consider how these ideas can be worked with to produce new knowledges and understandings of the city by considering one specific issue: the association between cities and creativity. While the book concludes by recognising the impossibility of developing an all-encompassing theory of the city, it ends by restating the case for taking cities seriously in human geography. As I argue in the conclusion – and throughout this book – geographers' persistent failure to clarify what is 'urban' about urban geography seriously impoverishes the discipline's comprehension of the relationship between society and space. As well as shedding light on different strands of urban theory, it is thus hoped that this book inspires geographers to consider new ways of thinking about cities and their endlessly fascinating spaces.

1

URBAN THEORY, MODERN AND POSTMODERN

Virtually all theories of the city are true, especially contradictory ones.

(Jencks 1996: 26)

In the introduction, I suggested that cities have frequently been identified by geographers as distinctive spaces. In turn, the distinctive nature of cities is deemed to require the development of specific ideas to describe and explain them. These ideas are collectively termed *urban theory*. However, the notion of theory is a complex one, and much misunderstood. Commonly, it is assumed that theory constitutes a series of big, abstract and grand ideas which help us make sense of the detailed complexity of everyday life. Conventional scientific methodology dictates that such theories can be constructed only on the basis of repeated and verifiable observation of what exists in the world. If these observations and measurements tally with the scientists' ideas of what is going on in the world, then these theories can be framed as 'laws' about how the world works. Urban theory might therefore be characterised as a set of explanations and laws which explain how cities are formed, how they function and how they change. Rather than being concerned with empirical nuance of city life (the daily nitty-gritty of urban existence), we might logically conclude that urban theory is preoccupied with grand themes – and is inevitably more airy-fairy in nature.

While this is a fair description of many urban theories, it is incorrect to suggest that all theories are of this type. In simple terms, a theory can be defined as an idea about what is going on in the world, and although many theories seek to connect a variety of empirical evidence to construct an explanation, theories need not be 'big'. Nor should we expect that all theories are empirically verifiable (i.e. can be shown to be true or false by reference to what we see happening in the world). As we shall see throughout this book, some theories are highly conjectural, and hypothesise the existence of things that cannot be shown to exist at the level of observable reality. Others are highly contingent and localised. Yet what all urban theories have in common is that they seek to make the city legible to us in some way or another. This might be through the publication of a book or academic paper outlining ideas about how the city works, a model of urban structure or the development of a computer programme that maps urban flow. It might even be through the production of a poem, a piece of art or a song designed to stimulate new understandings of the city.

Hence, although many important geographical texts on theory suggest that theories are connected statements which seek to explain geographical phenomena rather than merely describe them (see Johnston 1991), I want to start with a somewhat looser definition: a geographic theory is an attempt to think space in a new manner (see also Hubbard et al. 2002). Trevor Barnes (2001) offers a similar definition, suggesting that geographic theory need not provide logical relations, rules of causation or empirically verifiable statements. For him, what is crucial is that a theory *expresses* a phenomenon through a new vocabulary and syntax which changes the way people interact with it (in terms of how they view it, or study it, or practise it). Commonly, this involves taking ideas and concepts developed in one discipline to illuminate phenomena that lie outside their disciplinary remit. In human geography, for example, theoretical development has principally occurred through the deployment of languages and ideas developed in other disciplines, such as mathematics, physics, cultural studies, psychology, biology, politics, ergonomics, economics, sociology, literary studies, criminology and history. The fact that these disciplines cover the humanities, social sciences and natural sciences might be taken to indicate geography is a uniquely integrative discipline: in fact, all disciplines develop and evolve by drawing on other areas of knowledge as they search for new ways of thinking about their subjects and objects of study.

Given this definition of theory, my objective in this chapter – to consider the evolution of theories of the city – is potentially massive. It is rendered particularly problematic given the city has constituted an object of study for disciplines beyond geography (notably sociology, where urban sociology has become a recognised subdiscipline, albeit one whose identity has changed markedly over the years). Given this complexity, in this chapter I simply wish to review some of the best known and most influential theoretical frameworks which have been employed for making sense of cities. Here, I particularly focus on those theories that produced new languages and vocabularies that are now part of the established lexicon of urban geography. Even so, there are some notable absences and silences in this story, with many key thinkers and key ideas in the history of urban studies neglected. Some of these silences are redressed in later chapters, where some less celebrated urban theories are considered. Others are a consequence of the fact that this book is written for geography students in the English-speaking world, and thus makes scant reference to any writers whose work has not been made available in translation. In any case, readers are recommended to hunt down other, more fulsome, accounts of the development of urban theory, and to consider how urban geography fits within wider histories of the discipline (some suggestions for further reading are given at the end of the chapter).

MODERNITY AND URBANISATION

The idea that cities require particular diagnostic tools and conceptual languages can be understood only in relation to the emergence of cities themselves as distinct and recognisable phenomena. That said, there is little consensus as to when or where the first cities emerged: candidates include a number of regions in South West Asia, from the highlands of Mesopotamia through to the fertile valleys of Sumeria. Moreover, there is no clear agreement as to why cities first formed. Perhaps it was for defensive reasons; maybe it was a response to the emergence of a political or military elite, or a function of economic imperatives that required the creation of trading and commercial markets the type of which simply could not be accommodated in a village. Yet others believe that the search for a single explanation is futile, with several factors combining differently to encourage urban growth in different regions (see Pirenne 1925).

We return to some of these debates in later chapters. At this point, however, it will suffice to say that cities have been in existence for a long time. In this light, it is surprising that it took so long for a distinctive body of urban theory to emerge. Indeed, it was not until the nineteenth century that the city began to be taken seriously as a distinctive and important object of study. Perhaps the main reason for this is that until that time the share of the global population living in cities was relatively small. The rapid urbanisation of the nineteenth century (sometimes termed the second 'urban revolution') changed this, first taking root in the economically dominant states of the European heartland. For instance, through the 1800s, the United Kingdom experienced the most rapid and thorough urban-isation the world had ever witnessed. The 1851 Census of the Population of England and Wales revealed a watershed event: for the first time, more people lived in towns than in the countryside. Furthermore, 38 per cent of Britain's population lived in cities of 20,000 or more, and by 1881, this figure topped 50 per cent. From 1801 to 1911, the urban population increased nearly by a factor of ten, from 3.5 million to 32 million. Nowhere was this more evident than in the capital, London, which surpassed Beijing to become the world's largest city in 1825, and by 1900 had 6,480,000 occupants (2 million more than its nearest rival, New York).

It is quite conventional to suggest that the major transformation that triggered this rapid shift in settlement from the country to the city was the 'industrial revolution', which ushered in a world where manufacturing production was the driving force of societies. In industrial societies, the power of merchants, craftsmen and guilds was supplanted by the power of industrial capitalists making the commodities of an industrial age. These industrialists imposed their identity on the city in the form of a new landscape of industrial factories, workshops and machinery, around which were spun new webs of transportation and power. Both industrial employers and employees alike thus found themselves in an environment that was far removed from the rural settlements that predominated in feudal times: it was a city of social mobility, offering contrasts of wealth and deprivation from the spectacular residences of the nouveaux riches bourgeois to those street dwellers who relied on their wits to make a living among the detritus of urban life. It was also a city characterised by perpetual transformation, with new ideas, technologies and practices constantly transforming people's relation to one another as well as their place within a rapidly changing urban environment (see Plate 1.1). In short, it was a modern city:

Plate 1.1 Different social types share the newly constructed spaces of the modernising city – *A Spring Morning, Haverstock Hill* (George Clausen, 1881) (courtesy of Bury Art Gallery and Museum Archives)

There is a mode of vital experience – experience of space and time, of self and others, of life's possibilities and perils – that is shared by men and women all over the world today. I will call this body of experience 'modernity'. To be modern is to find ourselves in an environment that promises us adventure, power, joy, growth, transformation of ourselves and the world – and, at the same time, that threatens to destroy everything we have, everything we know, everything we are. Modern environments and experiences cut across all boundaries of geography and ethnicity, of class and nationality, of religion and ideology: in this sense, modernity can be said to unite all mankind. But it is a paradoxical unity, a unity of disunity; it pours us all into a maelstrom of perpetual disintegration and renewal, of struggle and contradiction, of ambiguity and anguish. To be modern is to be part of a universe in which, as Marx said, 'all that is solid melts into air'.

(Berman 1983: 1)

The ambivalent experience of space and time associated with the modern city was frequently juxtaposed with traditional ways of life regarded as socially more secure and predictable. However, such traditional ways of life seemed to persist in the rural, where the ties that bound family, land and labour remained strong.

Geography had yet to emerge as an institutional and academic discipline in the nineteenth century, save for the work of those explorers, cartographers and expeditionists who sought to chart the ways of life characteristic of distant lands. As such, the new urban forms characteristic of the nineteenth-century city were of little interest to geographers. In contrast, the discipline of sociology coalesced in the nineteenth century around questions concerning the differences between urban modernity and rural tradition. According to Savage et al. (2003), nineteenth-century sociologists began to be concerned with, inter alia, the following 'urban' questions:

1 Whether it was possible to talk about a distinctive urban way of life, and, moreover, to identify whether this was a way of life common to all cities.
2 Whether this urban way of life encouraged the formation of new social collectivities and identities (such as the formation of neighbourhoods).
3 Examining how urban life was affecting traditional social relations of deference, such as those in which someone's class position, gender, caste or race was seen as central.
4 Identifying how the city encouraged or discouraged the development of social bonds between citizens of different backgrounds, residence and occupation.
5 Documenting the history of urbanisation and explaining why populations were concentrating in towns, cities and conurbations.
6 Identifying the basic features of the spatial structure of cities and establishing whether different spatial arrangements generated distinctive modes of interaction.
7 Diagnosing 'urban' ills such as congestion, pollution, poverty, vagrancy, delinquency and thuggery.
8 Specifying how urban politics was emerging as a distinctive form of governance with specific and uneven impacts for different citizens.

Although some sociologists addressed these issues tangentially, and were more concerned with exploring the nature of modern life rather than urban life per se, the inevitable conflation of modernisation, urbanisation and industrialisation served to establish some important ideas about urbanism as a way of life. Foremost here was the idea that cities were larger, more crowded and socially mixed than rural settlements, bequeathing them a social character which was less coherent and more individualised than that evident in the rural.

This idea of the city as representing the antithesis of traditional ruralism is perhaps most associated with the work of the sociologist Ferdinand Tonnies (1887). In essence, Tonnies described two basic organising principles of human association. The first was the *gemeinschaft* community, characterised by people working together for the common good, united by ties of family (kinship) and neighbourhood and bound by a common language and folklore traditions. At the other end of the spectrum, Tonnies posited the existence of *gesellschaft* societies, characterised by rampant individualism and a concomitant lack of community cohesion. Though Tonnies couched the distinction between *gemeinschaft* and *gesellschaft* in terms of a pre-industrial/industrial divide rather than a rural/urban one, his description of *gesellschaft* societies was deemed apposite for industrial cities where the extended family unit was supplanted by 'nuclear' households in which individuals were first and foremost a unit of economic and social reproduction, fundamentally concerned with their own problems, and not those of others – even those in their immediate neighbourhood.

Of course, the idea that cities are bereft of community is a caricature, and it is worth underlining that Tonnies was describing two ends of a social continuum, not positing a binary distinction. Nonetheless, the idea that social cohesion and sense of community would be less evident in a city than the country was a persuasive one, and one which was to guide sociologists as they further explored the nature of modern urban living. Like Tonnies, many were clearly nostalgic for a rural way of life that seemed to be fast disappearing (this pro-rural sentiment is a recurring feature in much urban writing, and we will encounter examples later in the book). However, there were dissenters. For instance, one of the 'founding fathers' of sociology, Emile Durkheim developed a model of contrasting social order somewhat at odds with that of Tonnies. For Durkheim (1893), traditional, rural life offered a form of *mechanical solidarity* with social bonds based on common beliefs, custom, ritual, routines, and symbols. Social cohesion was thus based upon the likeness and similarities among individuals in a society. Durkheim argued that the emergence of city-state signalled a shift from mechanical to *organic solidarity*, with social bonds becoming based on specialisation and interdependence. Durkheim consequently suggested that although individuals may perform different tasks and often have different values and interests, the very survival of urban societies depends on the reliance of people on one another to perform specific tasks. Hence, in contrast to feudal and rural social orders, urban society was one which allowed for the coexistence of social differences, with a complex division of labour (where many different people specialise in many different occupations) creating greater freedom and choice for individuals. Optimistically, Durkheim's work therefore pointed to a new kind of solidarity, with people brought together by a new form of social cohesion based on mutual interdependence.

Irrespective of the emphasis that some sociologists put on the idea that modern cities were socially emancipatory, a more common prognosis was that cities were essentially cold, calculating and anonymous. Friedrich Engels' (1844) work is of particular note in this respect, given that it documented the inhuman living conditions experienced by workers in the industrialising metropolis (relating this to capitalist imperatives, as we will see in the next section). In a more general sense, the work of Georg Simmel (1858–1918) described the impacts of the city on social psychology, suggesting that the city required a series of human adaptations to cope with its size and complexity. In his much-cited essay 'The metropolis and mental life', for example, Simmel (1950) argued that the unique trait of the modern city was the 'intensification of nervous stimuli with which city dweller must cope'. Describing the contrast between the rural – where the rhythm of life and sensory imagery was slow – and the city – with its 'swift and continuous shift of external and internal stimuli' – Simmel (1950) thus detailed how individuals psychologically adapted to urban life. Most famously, he spoke of the development of a blasé attitude, which can perhaps be best described as the attitude of indifference which (most) urban dwellers adapt as they go about their day-to-day business. Simply put, to cope with the constant cacophony of noise, sights and smells that present themselves to us as we move through the city, Simmel suggested that we learn to cut out all stimuli which are not important to the business in hand.

Perhaps the most vivid illustration of the adoption of a blasé attitude is bystander indifference. This is the phenomenon whereby citizens will ignore an incident because it is none of 'their business'. The psychologist

Stanley Milgram ran a series of infamous experiments to illustrate this. These included asking his students to stage fights in urban public spaces, then measuring public responses. Remarkably, but perhaps not unexpectedly, the majority of local bystanders did not intervene, perhaps considering that there were plenty of other onlookers who could intervene and, in any case, they would rather not get involved (Milgram 1970). This abdication of responsibility was, however, not evident when these experiments were repeated in rural settings, where bystanders seemed to have much more of a sense of collective responsibility (and perhaps more of desire to maintain the 'order' of their rural locale). In other experiments, rural dwellers seemed much more willing to help when strangers in their midst requested help; again, in cities, it seemed urban dwellers felt that this was simply not their job. As such, the adoption of a blasé outlook and an indifference to one's urban surroundings may be a factor in the contrast between urban and rural crime rates: shockingly, there have been high-profile rape and assault cases in cities where bystanders have failed to intervene. Against this, villages are traditionally understood to be characterised by a higher level of natural surveillance and intervention.

The failure to challenge antisocial behaviour or to 'care' for our fellow citizens is just one symptom of the indifference which urban dwellers cultivate to cope with the complex demands of urban living. Another commonplace strategy of 'distancing' is to avoid encounters with strangers when we walk around our towns and cities, keeping contacts with other pedestrians brief and superficial. Given that most of the people we encounter in the city are strangers known to us only in categorical terms (e.g. fat, white, old, male), we make judgements about them simply by looking. This is so rapid it is subliminal (and seldom something we are conscious of): a quick flick of the eyes as people glance at each other, scan their appearance and then look away, having established that neither is dangerous. The streets of our cities are therefore a realm of what Lofland (1998) calls civil inattention, where people rapidly scan each other to gain some categorical knowledge about them, before turning their glance away for fear of invading their privacy. This etiquette maintains civil order (if you stare, this constitutes uncivil attention) and helps the citizen cope with the sheer number of strangers that are routinely encountered on the city's streets. In contrast, in less densely occupied villages, the fewer number of encounters means that the villager can expend more time and energy ascertaining the identity of a stranger to their village.

Of course, by the same token that glancing at others enables us to efficiently read their character and intention, we too are under surveillance. Merging into the urban scene is itself an acquired skill, as Erving Goffman (1959) famously detailed in The Presentation of Self in Everyday Life. This volume provided a detailed 'theatrical' description and analysis of process and meaning in street interactions, suggesting we are on a 'stage' when we are on the streets, constantly aware that we are presenting ourselves to an audience (of strangers). The process of establishing we are not a threat, then, becomes closely allied to the concept of the 'front', which may be different from the way we act and dress 'offstage' (e.g. in private space). This typically involves managing our bodies so they do not attract undue attention: in effect, we dress and act to send out a message – 'I'm normal, I'm not a threat!' Conforming to fashion thus becomes a way of preserving a sense of self, the equivalent of donning a social mask which identifies us as part of the (anonymous) crowd (Allen 2000a). Conforming to this prescribed notion of what is normal, rather than constantly seeking to express ourselves through our dress and actions, consequently becomes another way that we negotiate the everyday city and avoid social conflicts.

The idea that the urban experience is essentially 'managed' through a transformation of individual consciousness that involves a filtering out of the detail and minutiae of city existence is thus an important foundational story of modern urbanism. So too is the idea that city life debases human relations, and renders contact between urban dwellers essentially superficial, self-centred and shallow. For Simmel, the impersonality and lack of depth in urban life was fundamentally related to the fact that the industrial city was brought into being to serve the calculative imperatives of money. Simmel essentially suggested that the dominance of the money economy created interactions based purely on exchange value and productivity (and thus dissolved bonds constructed on the basis of blood, kinship or loyalty). This, he argued, encouraged a purely logical way of thinking which values punctuality, calculability and exactness (see Hubbard et al. 2002). Hence, while money fulfilled a multitude of tasks in the city by providing a medium of equivalence, the come-uppance was to reduce quality to quantity and to effectively destroy the essential 'form' and 'use' of any object encountered in the city: all was calculable. Simmel contended that this notion of calculation extended to other facets of urban life. One example was his suggestion that urbanites became highly attuned to clock time, contrasting this with the casual and vague sense of time predominant among rural dwellers who moved to the diurnal and seasonal rhythms of nature (see also Chapter 5 on the invention of modern time). For instance, Simmel hypothesised that if all the watches in Berlin suddenly went wrong, the city's entire economic and commercial life would be derailed. This underlines that the life of cities had become inconceivable without all of its relationships being organised and calculated within a firmly fixed framework of time (and space) which was designed to undermine individuality and spontaneity.

The idea that money undermined and 'hollowed out' individuality was something that Simmel's pupil Siegfried Kracauer was to elaborate in his writing on Weimar Berlin, a city that appeared relentlessly modern. As he put it, 'it appears as if the city had control of the magic means of eradicating memories . . . it is present day and makes a point of honour of being absolutely present day' (Kracauer 1927: 75). Suggesting that the aesthetic forms and spectacles of the modern city were a product of urbanisation and the modes of vital experience associated with urbanism, Kracauer thus explored the development of distinctly urban visual cultures. One example he cited was the development of shop window culture, with an excess of commodities leading to the phenomena of goods needing to have an 'enticing exterior' to distinguish them from other goods with equivalent or similar 'use value'. Shop goods thus needed to be fashioned, packaged and displayed in an aesthetic manner to increase their appeal, and shops too began to transform to allow for window-shopping. Yet Kracauer noted that the rise of 'surface culture' extended to all facets of the city: the design of housing, the aesthetics of the street, even the bodies of urban dwellers themselves. Noting the ludic pleasures of visually consuming the streets of modern cities, his work also emphasised the sheer duplicity of this process – which he suggested was deeply embedded in the emergent structures of the industrial capitalist system. For Kracauer, as for Simmel before him, reading the visual forms and commodities on display in the modern city thus revealed the paradoxes and possibilities of a distinctly modern – and inevitably urban – way of life. Summarising, his essay on the 'mass ornament' suggested that 'the position that an epoch occupies in the historical process can be determined more strikingly from an analysis of its inconspicuous surface-level expressions than from that epoch's judgements about itself' (Kracauer 1927: 78).

URBAN PATHOLOGIES AND THE INHUMAN METROPOLIS

Although sociologists specified a number of facets of urban life which demanded human adaptation – not least the development of a blasé attitude – above all else it was the anonymity of the city which was to be posited as its defining characteristic. Indeed, Max Weber's celebrated essay 'The city' eschewed any attempt to determine a minimum population size for an urban settlement, arguing that the city needed to be defined as 'a settlement of closely spaced dwellings which form a colony so extensive that the reciprocal personal acquaintance of the inhabitants, elsewhere characteristic of the neighbourhood, is lacking' (Weber 1922: 1212). The lack of identity was seen to have profoundly ambivalent consequences for urban dwellers. In the first instance, it was suggested the city offered a space of liberty and autonomy precisely because it allowed individuals to escape from the bonds of kinship and community that often condemned them to particular social roles or identities. The notion that cities could be emancipatory for women, for example, was often noted in literatures on urban sociology (though, conversely, this was also a source of anxiety for some male writers, as we shall see in Chapter 2). The idea that the individual could melt into the crowd, yet conversely use this sense of anonymity to develop a new sense of freedom, was a powerful motif in formative urban theorising, and was often used to explain why the city was go-ahead, innovative and civilised (while the rural remained, in the words of Karl Marx (1867), constrained by the 'idiocy' of isolation). In short, anonymity promoted individualism, free-thinking and civility, with the city creating the possibilities for meetings and mismeetings that would spark creativity. In some accounts, cities have therefore been deemed to be the cradle of civilisation because they provide democratic spaces (like the Roman fora or Greek agora) in which individuals could debate political matters (see Habermas 1989).

The idea that the anonymity of the city allows us to forge new identities unburdened by our histories and biographies thus emphasises the liberatory potential of the city. Yet, as Tonnies' work signalled, the flipside of this anonymity and freedom can be rampant insecurity and a profound sense of being alone even though we are surrounded by people. The term *anomie* is widely used to describe this form of alienation (Walmsley 1988). Originating in the work of Durkheim, anomie is an idea which expresses the dissolution of the moral and social certainties that were (allegedly) widespread in rural or *gemeinschaft* societies. In common manifestation, anomie describes a person's inability to either care for or identify with any of the number of people who surround them in the city. An extreme form of indifference, Durkheim (1893) suggested that anomie was expressed in the explosion of antisocial and criminal behaviour evident in cities. Simply put, anomie suggested that criminals were not merely unwilling, but unable to empathise with their victims.

Louis Wirth (1938) developed these ideas when he talked of the declining importance of primary social relations in urban centres, suggesting that anonymous and fleeting social relations become the norm. For Wirth, the importance of indirect and impersonal social relations increased in direct proportion to the urbanity of a settlement, which he described as a function of a settlement's population size, density and diversity. In his oft-cited words, a city may be conceived as 'a relatively large, dense, and permanent settlement of socially heterogeneous individuals' (Wirth 1938: 7). As he detailed, increasing any one of these, in isolation or combination, has important effects for the social life of a given settlement. For instance, he argued that increasing the number of inhabitants in a settlement beyond a certain limit affects the relationships between them and the character of the city:

Large numbers involve . . . a greater range of individual variation. Furthermore, the greater the number of individuals participating in a process of interaction, the greater is the potential differentiation between them. The personal traits, the occupations, the cultural life, and the ideas of the members of an urban community may, therefore, be expected to range between more widely separated poles than those of rural inhabitants . . . The bonds of kinship, of neighbourliness, and the sentiments arising out of living together for generations under a common folk tradition are likely to be absent or, at best, relatively weak in an aggregate the members of which have such diverse origins and backgrounds.

(Wirth 1938: 22)

Wirth did not imply that urban inhabitants have fewer acquaintances than rural inhabitants (indeed, the opposite is often true). Rather he inferred that city-dwellers meet one another in highly segmental roles, meaning that impersonal, superficial and transitory relations are the norm. His examples were diverse, but by way of one illustration he argued that although a city-dweller might go to a supermarket on a regular basis, it is rare to be served by the same cashier. As such, while we may exchange pleasantries with those who serve us, it is unlikely they are known to us, and the relationship is one structured entirely around a *functional* economic exchange.

Wirth's thesis – namely, that we are surrounded in cities by people whom we will only ever have impersonal relations with – implies cities are places where we rarely feel a sense of belonging and identity. In many ways, this is related to another manifestation of anomie as an urban condition: the rise of mental illnesses and neuroses. While the history of mental illness is a complex one (see especially Foucault 1967), the connection between the rise of cities and cases of mental illness has often been read as evidence of people's failure to adapt to city life. For instance, agoraphobia was first diagnosed among urban dwellers, not as a fear of open spaces, but a pathological fear of spaces where encounters with others could not be avoided (for example, standing in a crowd or standing in a line, being on a bridge, and travelling in a bus, train or automobile) (Vidler 1994). Likewise, neurasthenia, hysteria and claustrophobia were all imaginatively located by psychiatrists within a city which offered untold strains and stresses (Callard 1998). The cumulative impacts of noise, visual pollution and sensory bombardment have also been implicated in the demonstrated association between schizophrenia and inner city living (Walmsley 1988), while crowding and high-density urban living is widely understood to create a number of psychological reactions, including introversion and withdrawal. Phobias and anxiety accordingly came to be seen as characteristic of urban life (Vidler 1994), giving rise to the dictum that cities may be made by humans, but humans are not made for cities.

Unfortunately, the seminal work of early sociologists on the psychosocial economies of urban space was later to inspire some relatively unsophisticated theories of the impacts of environment on behaviour. In extreme cases, statistical associations were made between, for example, particular forms of housing and specific criminal behaviours (see, for example, Coleman 1985 on the design deficiencies of British council housing estates). Oddly, some of this echoed nineteenth-century theories of environmental determinism in which the physical environment was thought to shape the health and virtue of urban dwellers. As Driver (1988) outlines, in the same way that putrid water and dirt was said to issue unhealthy contagion in the form of invisible 'miasmas', squalid and crowded living quarters were said to be infected by moral miasmas. A plethora of pseudo-sociologists and social reformers thus began collecting statistics on the social conditions of different urban locales, most famously in the case of Charles Booth (1889), whose fastidious mapping of nineteenth-century London popularised the notion of a dangerous urban underclass in which 'there was greater development of the animal than of the intellectual or moral nature of man'. Though Booth developed an understanding of urban malaise in which poverty and inequality were emphasised, others collected evidence which pointed more straightforwardly to the urban conditions that produced immorality and vice. For example, Cherry (1988) outlines the influence of the American economist Henry George on Victorian city-fathers. His view – 'that the life of great cities is not the natural life of man . . . he must, under such conditions, deteriorate physically, mentally, morally' (George 1884: 203) – was to find much popularity among those who felt the close confinement and foul air of big cities was inimical to mental and spiritual fitness. In France, meanwhile, the obsessive social hygienist Parent-Duchatelet made constant allusions to the proximity of spaces of sex work and poor sewerage, forging a connection between physical and moral impurity that justified both the repression of prostitution and the creation of more commodious cities. Others argued that the immorality of the city was by no means restricted to the lower classes, with the decadent and moneyed urban elite being tempted into this world of vice and 'sexual disarray' (Cook 2003). For instance, the celebrated sexologist Krafft-Ebing postulated that large cities were the breeding ground of degenerate sensuality. His thesis was that the constant activity of urban life led to a nervous collapse and 'a giving into' the confusion of the city. As contemporaries of Krafft-Ebing contended, 'the problem lay with the city itself, whose anonymity, artificiality and rampant commercialism overstimulated the libido and distorted the balance of nature within and between the sexes' (Forel and Fetscher 1931: 91)

Hence, while the work of nascent urban sociologists including Durkheim, Simmel, Weber, Wirth and others sought to develop nuanced accounts of the psychosocial life of cities, it was the identification of particular pathological conditions of the city that was to have most contemporary influence. Indeed, the ideas developed by sociologists about urbanism came to legitimise a series of interventions in the urban environment, not least in the form of town planning movements that drew intellectual sustenance from social science. Simply put, the majority of twentieth-century plans for urban redevelopment grew out of a desire to counteract what were seen as the inhuman dimensions of the late-nineteenth-century city: lack of light and air, unsanitary and overcrowded conditions, congested circulation that demanded the opening up of narrow streets and the reinvention of 'community' in the form of planned neighbourhood units and collective facilities. Some, like Le Corbusier, reimagined the city in glass and concrete, its transparent towers majestically spaced in a vast urban park (Fishman 1977). Others, like Frank Lloyd Wright, envisioned it dispersed in 'broadacres' spread across the fertile prairie, each citizen given a plot of land in a garden city. What all shared was an awareness of the ills of urban living and a desire to develop a form of urban living more suited to human needs and desires.

COMPETING FOR SPACE: URBAN ECOLOGIES AND LAND USE

The rapid urbanisation of the late nineteenth and early twentieth centuries clearly led to some important theories about the nature of modern urban living. It is difficult to argue that these ideas amounted to a coherent or overarching theory of the city; further, the city as an object of study remained tangential to most academic disciplines. However, the establishment of the first US Department of Sociology at Chicago in 1913 effectively established urban studies as a legitimate and important field of study. Its co-founders, Robert E. Park and Ernest Burgess, regarded Chicago as an 'urban laboratory' in which they could explore how humans adapt to the city. Crucial to the work of the so-called Chicago School (Box 1.1) was the idea that people – like other species – need to compete to survive. Robert Park (once a pupil of Georg Simmel) thus proposed the concept of human ecology as a perspective applying biological processes to the social world. This perspective advances the idea that 'cities are the outward manifestation of processes of spatial competition and adaptation by social groups which correspond to the ecological struggle for environmental adaptation found in nature' (Cooke 1983: 133). Although now largely discredited (Gottdiener and Budd 2005), the idea that there is a similarity between the organic and social worlds provided an important underpinning for the Chicago School's fastidious explorations of the urban scene, not least the many descriptions they completed of life in different communities and neighbourhoods.

Box 1.1 THE CHICAGO SCHOOL

Generally regarded as the first sociology department to be established anywhere in the world, the University of Chicago Department of Anthropology and Sociology became synonymous with urban sociology during the first half of the twentieth century. Although probably best known to geographers as the source of innumerable models of urban social structure (including Burgess's infamous and often-criticised concentric zones model), the work of the Chicago School was more diffuse and complex than is often acknowledged nowadays. For example, although the writings of its most famous members, W.I. Thomas (1863–1947), Robert E. Park (1864–1944), Ernest Burgess (1886–1966) and Louis Wirth (1897–1952), drew in an often spurious manner on concepts derived from eugenics and social Darwinism, they were based on fastidious empirical research involving intensive qualitative fieldwork and observation. For instance, Park et al. (1925: 37) wrote of the need to observe and document the characteristic types of social organisation and individual behaviours evident 'in Bohemia, the halfworld, the red-light district and other moral regions less pronounced in character', positing the marginal spaces of the city as 'laboratories' in which 'human nature and social processes may be conveniently and profitably studied'. Following Park's suggestion, members of the Chicago School were subsequently responsible for some remarkable *micro-sociological* descriptions of the customs and social practices of those found in marginal spaces, from Wirth's (1928) studies of the African American inhabitants of the 'ghetto' to Walker C. Reckless's (1926) descriptions of street prostitution. The Chicago School's advocacy of participant observation and *ethnographic* technique (literally, 'writing about a way of life') has remained extremely influential in urban sociology, particularly among those researchers who seek to shed light on the different social groups – or *subcultures* – existing on the margins of society (Gelder and Thornton 1997).

Further reading: Yeates (2001)

The influence of Darwinian evolutionary ideas was very evident in the work of the Chicago School. The suggestion that people may be defined as individual biological units involved in struggle for scare resources provided a (then) unique take on city life, postulating that cities are driven by competition, the structure of the city being an outcome of this natural competition (or symbiosis). In particular, it suggested that the most successful urban dwellers would take over the 'best' areas of the city to guarantee their 'survival', while the least successful would end up in less salubrious areas. This process of social competition thus bequeathed a spatial sorting, typically peripheralising the least successful and wealthy urban dwellers. Yet the Chicago School also focused on the 'biotic' ways in which citizens cooperated to survive in the city, with successive waves (or invasions) of in-migrants clustering together for mutual support and defence. Taken together, these ideas of 'natural' competition and cooperation suggested that the city could be described as a complex 'super-organism' containing 'natural' areas of many types: ethnic enclaves, activity-related areas (business, shopping, manufacturing etc.), differently classed housing areas (upper-class suburbs and working-class tenements) as well as spaces of sexual and social immorality (the city's 'twilight' zone).

Hence, while some of the pioneering writers in urban sociology sought to isolate particular characteristics of the urban condition – often diagnosing the specific pathologies of urban living – in the 1920s, Chicago became the centre of an alternative tradition of urban analysis which focused on city ecologies. Nonetheless, the vocabulary adopted by the Chicago School was incredibly broad, taking in market economics as well as Darwinian theories of competitiveness. Likewise, the Chicago School based their ideas on a range of research methods, which included ethnographic methods of participant observation alongside more established 'scientific' methods. The Chicago School were all, to a lesser or greater extent, academic mavericks, and the willingness of Park, Burgess, Reckless, Shaw, MacKay and others to immerse themselves in the lives of those who might otherwise have remained a mere footnote in urban studies (the prostitute, the 'hobo', gang members and so on) attracted much comment. In many ways, this form of ethnographic engagement with the life-worlds of urban dwellers was a key influence on the social geographers who later sought to develop a people-centred humanistic geography (P. Jackson 1985; Herbert 2000). In a similar sense, these studies also represented the first tentative steps towards a sociology of subcultures: for instance, their focus on the city as a sexual laboratory pre-empted more contemporary ethnographies of sexual culture and identity (Heap 2003).

Latterly, however, the fecundity of the Chicago School has often been reduced to a rather banal regurgitation of their ideas, and a brief dismissal of the biological concepts which underpinned their explorations. Often, it seems that the chief legacy of the Chicago School was the urban model – a simplified representation of the city which identified key social divides in the city in diagrammatic form. Most famous here, perhaps, is Robert Burgess's concentric zone model, a neat recapitulation of his theory that as cities grew spatially, patterns of land use would reflect the successive phases of invasion and occupation. The resulting pattern was one of a centralised business core surrounded by four principal zones (a zone in transition where newly arrived migrants would seek lodging, a zone of working-class housing, a settled residential zone and a ring of commuter suburbs). Although a description of the social segregation of Chicago, this diagram was often taken to be indicative of the type of social patterns expected in all industrial cities. The fact that it offered only a broad approximation led others (including Burgess's colleagues) to offer their own refinements. Most notable were Homer Hoyt's sector model of the city (which implied that transport corridors would produce patterns of intense competition for land and real estate in a way that Burgess's model did not predict) and Chauncy Harris and Edward Ullman's multiple nuclei model (which suggested there would be numerous foci for business investment rather than intense competition for the central business district (CBD)).

Urban models of this type were later criticised for attempting to shoehorn a complex social reality into a grossly simplified model which drew lines that didn't really exist around areas that didn't really matter. Perhaps these criticisms are misplaced if we view models (like any theory of the city) as an attempt to think about the city in a different way, employing a different (and in this case, pictorial) language. Further, these models have clearly had much use as pedagogical devices, and have acted as the springboard for hundreds – possibly thousands – of student studies, dissertations and theses. As such, these models have become totemic in urban studies, and stand as an important reference point in urban geography (so much so that it is nearly impossible to imagine an urban geography text that does not at least acknowledge their existence). What is particularly significant about these models is that they also represent one of the first attempts to draw general conclusions about the segregation and inter-urban differentiation of the urban landscape.

In the post-war period, as geographers began to draw inspiration from locational theory in economics, this preoccupation with mapping and modelling patterns of urban land use became an important tradition, and one of the clearest manifestations of geographers' desire to restyle their discipline as a *spatial science*. Herein, the language of biology was supplanted by new languages and concepts – not least the formal language of mathematics and statistics – which held out the promise of more precise and accurate models of urban process. The potential of new technologies (e.g. calculators, computers) to identify statistical regularities and patterns in quantitative data also provided the promise of producing models which were the summation of tested and verifiable hypotheses. Underpinned by positivist principles (see Hubbard et al. 2002), spatial science was an attempt to distance urban studies from the arts and humanities, and ally it with the 'hard' sciences by adopting the principles and practices of scientific method.

The promise of spatial science was to suggest that both human and physical geographers alike could enact a rigorous exploration of spatial structure – an argument developed in David Harvey's (1969) *Explanation in Geography*, which was perhaps the most complete theoretical statement of geography's quantitative ambition. By adapting and rewriting classical locational theory (especially the models of land use proposed by Von Thünen, Christaller, Weber and Lösch), Harvey and others proposed that there could be an integrative and comprehensive foundation for modelling geographical pattern and process. This suggested that the place of things could be mapped, explained and predicted through the identification of underlying laws – often mathematically derived – of interaction and move-ment, with friction of distance regarded as a key factor explaining patterns of human behaviour. For such reasons, many spatial interaction models were referred to as gravity models because they utilised Newtonian theories (of gravitational attraction) to model flows between (nodal) points. Thus, where a 'natural' science like physics tried to create general laws and rules about things like molecular structure, geography sought to create models of spatial structure which could, for example, generalise settlement patterns, urban growth or agricultural land use (e.g. Berry 1967). The analogous use of scientific theory even led some to propose that human geography could be described as spatial physics (Hill 1981), bequeathing it the status of a 'hard' scientific discipline rather than a 'soft' artistic pursuit. This seductive type of argument was typical of the case made for scientific human geography, with the standards of precision, rigour and accuracy evident in mainstream science proposed as the only genuinely explanatory framework available for the generation of valid and reliable knowledge (A. Wilson 1972). This was also a key factor encouraging the adoption of the ideas and language (if not always the method) of science among those preoccupied with the status of the discipline and the links between physical and human geography. Additionally, for those believing that geography should be engaging with policy debates, scientific geography appeared to have considerable potential to become 'applied' geography, offering an objective and value-free perspective on the success of, for example, urban planning policies (Pacione 1999).

The new concepts and methods of spatial science thus allowed for the refinement of ecological modelling, with data relating to land use, migration and rental values used to approximate more and more complex models of urban land use and social organisation. For instance, statistical methods of exploratory data analysis could be used to simplify vast amounts of data relating to different census tracts and neighbourhoods to identify related areas (Berry and Rees 1969). Factorial analysis in particular became a widely adopted method for crunching census data and identifying the social, economic and demographic variables most important in differentiating neighbourhoods (following the methods spelt out by Shevky and Bell 1955). Though later critiqued for being socially acquiescent and irrelevant to the problems of the modern city (Harvey 1973), such statistical procedures shed light on the spatial imprints of racism (particularly in the United States, where the ghettoisation of non-white urban residents was regarded as a consequence of 'white flight').

Typically, the city mapped and modelled via such methods was a modern industrial one, where residential segregation largely reflected the socio-economic status of residents, and where factors of family status and ethnicity remained subordinate to that of class. Likewise, it was also assumed to be a monocentric city where land values declined away from the central business district where competition for land remained highest. For instance, Alonso's (1964) much-cited model of urban land use effectively refined Von Thünen's model of agricultural land use to provide a theory of how the urban land market organised around the city's CBD. His theory aimed to describe and explain the residential location behaviour of individual households and the resulting spatial structure of an urban area. The central concept of this theory was bid-rent, defined as the maximum rent that can be paid for a unit of land (e.g. per hectare) some distance from the city centre if the household is to maintain a given level of utility. Assuming jobs and retailing were concentrated in the city centre, the model implies that the price of both housing and of commuting depends on distance from the city centre, with a distance decay relationship between land rent and distance from the CBD. This suggests that the further a household lives from the city centre, the more it will have to spend on commuting and the less it will be able to spend on housing. The bid-rent curve of the actual land rents in the city is accordingly interpreted as reflecting the outcome of a bidding process by which land is allocated to competing uses (e.g. residential versus commercial uses). However, as in von Thünen's original model of agricultural land use, a monocentric, flat, continuous and uniform urban area was assumed. Based on these assumptions, bid-rent curves were assumed to be downward sloping (i.e. rent decreases with distance from the city centre to compensate for commuting costs) and single valued (so that for a given distance from the CBD only one rent bid is associated with a given level of utility).

Retrospectively, it has been suggested that the assumptions under-pinning bid-rent models limit their usefulness in approximating observed land-use patterns or analysing land-use change. Likewise, while statistical analysis of objective social facts (e.g. the proportion of non-white residents in an area, the number of crimes, the percentage of home-owners) allowed for the construction of elegant descriptions of social segregation, some commentators suggested these could never approxi-mate reality until they accounted for the variability and irrationality of human behaviour. Essentially, spatial science assumed that human actors were rational men (sic), fully aware of variations in land markets and able to judge the optimum location in the city they could afford. In practice, however, people often live in suboptimal locations because they lack perfect knowledge or ability to act on that knowledge. This means that the logic of the market does not hold, and that the patterns of residential segregation and land use hypothesised by abstract urban models are often only weakly predictive.

Acknowledging these weaknesses, one outgrowth of positivist urban studies was a 'behavioural turn' which sought to explore how people act on the basis of distorted, simplified and biased knowledge of their surroundings. Hugely influential here was the work of Kevin Lynch (1960) on *The Image of the City* – a book which introduced the notion of the mental map. This is the cognitive or mental picture of a city we carry in our heads and refer to when we make decisions about how to get from one area of the city to another, for example, or decide where we want to rent a house. Far from being an accurate and scale map of the city, this is a map full of contradictions and inconsistencies, in which 'real' distances are distorted and routes misplaced. Although Lynch's initial project was an attempt to show planners and architects how to design legible cities, it was to inspire a legion of studies which attempted to see how people thought about and acted in their cities, particularly in relation to topics such as spatial way-finding, navigation and cognitive distance estimation (see Holloway and Hubbard 2001). Yet it also suggested there might be possibilities for strengthening urban models by including variables that accounted for the variability and irrationality of much human behaviour. For instance, models such as Alonso's were based on ideas that householders possessed perfect knowledge of the actual land rent structure in a city, choosing a location that maximises their utility subject to their budgetary constraints; behavioural studies suggested this perfect knowledge was simply not present, and that suboptimal location decisions were the norm.

A strand of empirical work thus emerged which eschewed analysis of objective indicators in favour of exploration of people's subjective understanding of the city and its land markets. In many cases, this involved an attempt to incorporate people's decision-making processes into urban models, rejecting utility theory in favour of a more nuanced account of people's preferences as revealed through questionnaires. Such studies demonstrated that housing choice could not be regarded as rational, and the process of searching for and evaluating housing is unsystematic and normally based on imperfect information. In other studies, housing preferences were merely inferred from actual behaviour, with data on tenure choices and commuting patterns used to construct models based on random utility theory or entropy maximisation (Thrift 1980). Work on people's urban activity spaces and travel behaviour also encouraged experimentation with computer simulation, gaming approaches and fuzzy set theory, each of which provided the promise of recognising the contradictory and ambiguous nature of urban decision-making. Although such models still aimed to simplify the messiness of urban life through quantification and data reduction, their focus on the spatial preferences and behaviours of individuals provided an important rejoinder to those who suggested that urban modelling abstracted individuals 'out of existence'.

In these different ways, urban geographers began to flesh out the internal dynamics of the city and consider the impacts of human spatial behaviour on the urban landscape. Yet at the same time, geographers were also interested in the external relations of cities, borrowing from the central place theories emerging in economic geography (via Walter Christaller). Generalisations of these studies linked with formal models to suggest that intercity relations were organised as national urban systems forming urban hierarchies in which smaller cities had lower-level urban functions. The two sides of this urbanism were famously brought together by Berry (1964) in his account of 'cities as systems within systems of cities'. But no sooner had this neat arrangement been codified and widely disseminated through urban geography textbooks than both urban patterns began to alter: processes of deindustrialisation recast the relationship between the city and its suburbs, while processes of economic globalisation meant that intercity relations fundamentally changed. Further, by the 1970s there were other currents in human geography that were leading some to question the relevance of abstract empiricism; namely, the sustained challenge of Marxist and radical ideas.

MARXISM: A RADICAL INTERVENTION

Although it would be wrong to suggest that the 'ecological model' was atheoretical (Knox and Pinch 2001: 216), its brand of abstracted empiricism came under sustained critique in the 1970s as theories which emphasised social conflict in the city came to the fore. Crucial here were new forms of urban analysis which, conversely, borrowed from the long-standing sociological perspectives of Weberian and Marxist theory (Castells 1977; Harloe 1977; Pahl 1977; Saunders 1981). In relation to the former, Weber's ideas about key individuals influencing the distribution of social goods were to prove significant in the development of urban managerialism, a perspective which explored how key social actors controlled urban assets and land markets. Such actors included housing managers, planners, estate agents, mortgage lenders, financiers, police, councillors and architects. Collectively and individually, it was argued these actors could deny certain social groups access to particular property markets (and hence, particular parts of the city). Famously, Rex and Moore (1974) demonstrated that these actors often sustained racial divides in the city by steering non-white groups to particular areas (so that, for example, estate agents could encourage certain social groups to locate in some areas, while council workers might 'dump' ethnic minorities in certain hard-to-let housing estates). These descriptions of institutionalised racism and discrimination highlighted the forms of social inequality that were writ large in the urban landscape, and dispelled any notion that housing and land markets were in any way open to free competition (contradicting many positivist urban models).

This focus on the role of managers in influencing the distribution of urban resources was to prove pivotal in the emergence of a 'new urban sociology'. In the first edition of the acclaimed *Whose City?*, Pahl (1970) concluded

a truly urban sociology should be concerned with the social and spatial constraints on access to scarce urban resources and facilities as dependent variables and managers or controllers of the urban system, which I take as the independent variable.

(Pahl 1970: 221)

The new urban sociology was thus concerned with questions of conflict and injustice, poverty and racism, issues that appeared to have been neglected in previous studies. Here, it is important to acknowledge the political and social changes that were occurring at this time: 1968 had witnessed student riots in Paris, there was a growing awareness of sexism throughout the West, and homophobia and incidences of racial intimidation and violence were widespread. Against this backdrop, many commentators began to question the relevance of spatial science to urban social problems (e.g. Harvey 1973; D.M. Smith 1975). For these writers, urban studies at the time appeared to be populated by practitioners who were constructing models and theories in splendid ignorance of the problems of those living in the world beyond the 'ivory towers' of academia. Ironically, it was one of the most forthright proponents of quantitative geography – David Harvey – who now sought to propose a radical Marxist geography. In his oft-cited words:

There is a clear disparity between the sophisticated theoretical and methodological frameworks which we have developed and our ability to say anything really meaningful about events as they unfold around us . . . There is an ecological problem, an urban problem, a debt problem, yet we seem incapable of saying anything in depth or profundity about any of them.

(Harvey 1973: 129)

In this regard, the change in the theoretical orientation of urban studies stemmed from the general disillusion among urban scholars concerning the inability of dominant approaches to give a satisfactory explanation for the forms of inequality occurring in cities.

The urban sociology that emerged in this period accordingly made great play of its social commitment. Accordingly, Milicevic (2001) describes the characteristics of the new urban sociology as being its criticism of existing urban sociology and reinterpretation of concepts like urban, urbanism and urbanisation; its emphasis on relationships of production, consumption, distribution, exchange and power; its designation of social conflict and change as issues of special importance; and its concern with patterns of exclusion and inequality. While some of those working in the positivist tradition shared some of these concerns, there were sharp differences in terms of the relationships that 'new' urban sociologists forged with urban policy-makers. Previously, it had often been imagined that the role of researchers was to provide policy-makers with useful information – a position regarded with suspicion by those who now regarded institutions such as planning authorities and housing agencies as major factors in the production and reproduction of social problems. As such, the new urban sociology adopted a more confrontational stance to those in authority (and many of its practitioners were active in organising grassroots activity and protest).

In sum, urban managerialism dismissed positivist urban scholarship as theoretically barren and politically acquiescent. By exploring the role of key gatekeepers in allocating resources, managerialism highlighted some of the distinctive forms of social conflict played out in the urban realm. Yet managerialism was shortly to be overshadowed by a more fundamental form of critique – that of Marxist urbanism (see Box 1.2). At the heart of Marxist theory is the idea that society is structured by transformations in the political economy, and is organised so as to reproduce specific modes of production (such as feudalism, capitalism or socialism). Most significant in the context of urban writing is the capitalist mode of

Box 1.2 MARXIST URBANISM

Karl Marx remains one of the most widely discussed and written-about figures in academia over 150 years since his major works – the unfinished volumes of *Das Kapital*, *The Communist Manifesto* and *Grundrisse* ('Outline of a Critique of Political Economy') – were published. For some urbanists, his ideas remain inspirational, offering both an accurate description of the processes that drive capitalist cities and a set of prescriptions for the injustices and inequalities associated with that city. For others, he serves as a 'straw man' whose thinking on political change failed to predict the injustices that could be served in the name of socialist progress and class revolution (as witnessed, for example, in Stalinist Russia) and whose major legacy is to have perpetuated a dogmatic and inflexible way of thinking about social life. It is difficult to reject either set of arguments; Marx's ideas were a product of their times and his emphasis on class relations was an obvious response to the changes occurring in industrialised capitalist cities (particularly Britain) in the nineteenth century. Yet the idea that the organisation of space is fundamental in the reproduction of labour power (and hence capitalism) provided a distinctive take on the city – one developed by subsequent generations of urban researchers (notably Lefebvre, Berman, Castells, Harvey and Debord – see Merrifield 2002). What is perhaps most interesting about Marx's ideas is that, over 150 years after first being formulated, they still provide a framework for analysing social process. Above all else, it is the failure of Marx's classless 'self-determined' society to have materialised that actually inspires many social scientists to continue to explore his ideas. In relation to urban studies, this is evident in the work of those who have explored the role of the city in reconciling and diffusing the contradictions of capitalism: to paraphrase Lefebvre (1991), we do not know why capitalism continues to survive, but we know how – by *occupying place* and *producing space*. Marx's legacy for urban studies is his insistence on a structural reading of society and his careful articulation of capitalist process; even in the present, when the reductionism

continued

of his ideas is derided by many, questions of capital and class remain prominent in radical and critical geographical writing on the city.

Further reading: Harvey (1982); Castree (1999); Blunt and Wills (2000)

production, which, according to Karl Marx, first developed when labour itself became a commodity (i.e. when peasants became free to sell their capacity to work). In return for selling their labour power, such labourers received money which allowed them to survive. Marx described those who sell their labour power to live as *proletarians*. Conversely, the person who buys labour power (and typically owns the land and technology required to produce goods) was described as a member of the *bourgeois*. Marx's ideas – primarily developed in the context of nineteenth-century industrialisation – suggested that the bourgeois class took advantage of the difference between the price of the labour required to produce commodities and the value they could obtain in the marketplace. Here, Marx observed that in practically every successful industry the price for labour was lower than the price of the manufactured good (its 'exchange value'). Marx termed this difference 'surplus value' and argued that this surplus acted as the source of a capitalist's profit. One important side-effect is that social relationships between people are reconstituted as relationships between things, with people obliged to become consumers and purchase products they and others have made in the workplace. Given that exchange values have supplanted 'use' values over time, he argued it would be inevitable that workers become alienated or estranged from the products of their labour, with spatial and social division between production and consumption resulting.

In Marx's account, the capitalist mode of production is capable of tremendous growth because the bourgeoisie can reinvest profits in new technologies and produce new commodities. But Marx believed that capitalism was prone to periodic crises, such as technological obsolescence, over-production and falling demand for commodities. Consequently, he suggested that over time, capitalists would invest more and more in new technologies, and less in labour. Since Marx believed that surplus value appropriated from labour is the source of profits, he concluded that the rate of profit would fall even as the economy grew. When the rate of profit falls below a certain point, the result would be a recession in which the economy would collapse. Marx believed that this cycle of growth, collapse and growth would be punctuated by increasingly severe crises (ultimately fuelling class revolt).

Although Marx said little explicitly about cities, he did acknowledge the importance of space in overcoming these periodic crises and averting economic meltdown. Famously, he recognised that the 'annihilation of space by time' was fundamental in allowing capitalists to exploit new markets and populations. One facet of this was the growth of cities themselves, with Marx referring to towns both as concentrations of population and 'instruments of production'. Bringing workers together into pliant pools of exploitable labour, cities also provided a ready market for new commodities. But while increasing city size had advantages for the capitalist classes, Marx alleged that it resulted in the increased impoverishment of the proletariat:

> The more rapidly capital accumulates in an industrial town . . . the more miserable and impoverished are the dwellings of the workers . . . improvements of towns, such as the demolition of badly built districts, the widening of city streets, the erections of palaces to house banks or warehouses obviously drive the poor into even worse and more crowded corners.
>
> (Marx 1867: 65)

Yet it was Marx's colleague, Friedrich Engels (1844), who provided a clearer statement of the connections between city life and capitalism in The Condition of the Working-Class in England. This offered a damning indictment of the industrial capitalist city, based on his experiences in Manchester:

> If any one wishes to see in how little space a human being can move, how little air – and such air! – he can breathe, how little of civilisation he may share and yet live, it is only necessary to travel hither. True, this is the Old Town, and the people of Manchester emphasise the fact whenever any one mentions to them the frightful condition of this Hell upon Earth; but what does that prove? Everything which here arouses horror and indignation is of recent origin, belongs to the industrial epoch.
>
> (Engels 1844: 67)

Elaborating on the dialectic of industrialisation and urbanisation, Engels suggested that industry needed pools of dispensable labour to call on or cast off, and that the dense concentration of the working classes in areas of abject housing was the corollary of capitalist endeavour. He was also fiercely critical of urban renewal programmes where capitalists sought to improve working-class housing – dismissing these as attempts to keep housing problems in check, but not solving the housing problem. As such, Engels did not see a 'housing solution', and argued that only the end of capitalism would improve the condition of the working class.

In the work of Marx and Engels, dialectical thinking was therefore used to identify the forms of contradiction and tension in capitalist societies that needed to be resolved by transformations and adjustments to the mode of production. For instance, coming to recognise the fact that exploitation of labour power by the capitalist classes threatened to instigate class revolt, Marx wrote of the perpetual modernisation and agitation employed by the bourgeoisie to ensure that the relations of production were maintained. In modern, capitalist societies where 'all that is solid melts into air', Marx thus argued that new forms of socio-spatial relation were being constantly brought into being, reproducing capitalism (see also Berman 1983). Yet despite providing an intuitively attractive framework for exploring urban transformation, Marxist ideas remained largely unexplored by geographers and urbanists until the 1970s, when the search for more socially relevant knowledge and approaches alighted on Marxist theories. In effect, Marxism provided a revolutionary urban theory for revolutionary times, shifting the focus of urban studies from quasi-biological metaphors and mathematical models of urban process to analysis of the political and economic underpinnings of the urban system.

Inspired by Marx, Engels and others in the 'Marxist' canon (notably Althusser), urban scholars thus embarked on a new era of urban exploration in which the city was deemed to be a vital ingredient in a wider story of class conflict and ideological control. In the writing of those geographers and urban theorists who most directly engaged with Marxist theory – notably, David Harvey and Manuel Castells – this type of reasoning was transformed into explications of the role of space in this process of legitimisation and crisis avoidance. Simply put, these writers emphasised the importance of capitalism's *spatial fix* – the way that spatial differentiation and de-differentiation were implicated in capitalist relations. Allied with this focus was a renewed interest in the role of social agency. Initially, this was apparent in a number of studies of urban class consciousness, but latterly this bequeathed explorations of the role of the urban landscape in ideologically legitimising and celebrating the capitalist system (Debord 1967) (see also Chapter 2 on the 'duplicity' of landscape).

Perhaps the most significant proponent of Marxism in the new urban sociology was Manuel Castells. In *The Urban Question*, Castells (1977) extended Marxist perspectives on the city by exploring the dual role of the city both as a unit of production and a locus of social reproduction. The means whereby such reproduction was realised, he proposed, was *consumption* – the individual private consumption of food, clothing, as well as the collective consumption of such items as housing and hospitals, social services, schools, leisure facilities and so on. The implication here was that cities were organised so that the state (allied with capitalists) could provide the collective facilities needed to reproduce and maintain a flexible, educated and healthy workforce at least cost. This included the provision of state-subsidised housing (council housing), often organised in neighbourhood units centred on shops, community centres, doctors' surgeries and other communal facilities. This perspective stressed that the spatial form of the city was implicated in a number of significant ways in the reproduction of capitalism. For Castells, this reinforced the idea that 'spatial transformation must be understood in the broader context of social transformation: space does not reflect society, it expresses it, it is a fundamental dimension of society' (Castells 2000: 393). Castells was thus scathing of those commentators who seemingly 'fetishised' the urban, bequeathing it a distinct ecology that was somehow independent of capitalist structures. Developing this point, he outlined the need for a *structural* reading of the city:

> It is a question of going beyond the description of mechanisms of interactions between activities and locations, in order to discover the structural laws of the production and functioning of the spatial forms studied . . . There is no specific theory of space, but quite simply a deployment and specification of the theory of social structure, in order to account for the characteristics of the particular social form, space, and its articulation with other historically given, forms and processes.
>
> (Castells 1977: 124)

This structural solution to the 'urban question' thus offered a valuable corrective to the notions of human agency widely evident in urban studies at this time, whereby urban spaces were seen to be shaped by the knowledge and action of those who inhabited them. Yet, in offering this corrective, Castells seemingly went to the other extreme: space simply became a reflection of social process (hence, his surprising claim that 'space, like time, is a physical quantity that tells us nothing about social relations' – Castells 1977: 442). This is mirrored in Castells' definition of the city as 'a residential unit of labour power, a unit of collective consumption corresponding "more or less" to the daily organization of a section of labour power' (Castells 1977: 148). In this sense, the city was interpreted as the outcome of the state's provision of collective means of consuming commodities, something Castells felt could not be assured by capital but was nonetheless essential to the reproduction of capital.

In effect, Castells' radical take on the urban question shook up urban studies through its insistence that the social processes resulting in the production of the city were not distinctly urban, but endemic to capitalist society. This perspective was to be elaborated and extended by subsequent commentators, including David Harvey, whose various works provided a sustained and critical engagement with Marx's oeuvre. Identifying urban space as an *active moment* – a unit of capital accumulation as well as a site of class struggle – Harvey (1982) focused on the idea that surplus profit can be used to make more goods for short-term gain (primary circuit) or invested in property (secondary circuit) for longer-term gain, suggesting that in times of economic slump, profits could be most usefully ploughed into property – often triggering major urban renewal. This provided a different take on the urbanisation of capital, stressing the role of the built environment as a source of profit and loss:

> Under capitalism there is a perpetual struggle in which capital builds a physical landscape appropriate to its own condition at a particular moment in time, only to have to destroy it, usually in the course of a crisis, at a subsequent point in time. The temporal and geographical ebb and flow of investment in the built environment can be understood only in the terms of such a process. The effects of the internal contradictions of capitalism, when projected into the specific context of fixed and immobile investment in the built environment, are thus writ large in the historical geography of the landscape that results.
>
> (Harvey 1973: 124)

Harvey thus characterised the urban landscape as subject to contradictory impulses of investment and disinvestment, noting there is also an imperative to segregate upper-class and working-class residential areas to suppress working-class agitation. This segregation simultaneously created an urban landscape characterised by high and low land values. In particular, the association of specific areas with the 'underclass' drove down land prices in those locales, meaning that subsequent development could realise the difference between actual ground rent and the potential rent offered by that site (the so-called 'rent gap' – N. Smith 1979). Uneven development was accordingly theorised by Harvey as a crucial means by which capitalism could create for itself new opportunities for capitalist accumulation.

Harvey worked these ideas through in a series of celebrated accounts of nineteenth-century Parisian modernisation (summarised in Harvey 2003). Others focused on more contemporary processes of suburbanisation and urban restructuring. For instance, Allen Scott (1980) explored the *urban land nexus* in southern California, noting the differential locational advantages associated with different sites. In Scott's account, urban processes were conceived (*à la* Harvey) as involving the resolution of conflicts between capital and labour. Here, the local state was identified as significant, assisting capital accumulation through the provision of welfare and subsidies to business. Emphasising the importance of capitalist enterprises, Scott argued that urban development was a function of changing 'capital to labour' ratios among firms as they modified their production methods to maximise profits. One dominant trend (in southern California, at least) was the increased decentralisation of firms from the urban core – a movement enabled by technological and communication innovations. These changes in location of production space encouraged suburbanisation as households sought suburban locations closer to employment centres. It was here that the role of the state in assuaging the contradictions of capitalism was deemed crucial. In effect, the state became involved in 'unravelling the spatial knots' which this process of suburbanisation created, especially in relation to the provision of housing and transport infrastructure. As part of the state apparatus, urban planners were regarded as 'bees in the capitalist hive', working ceaselessly to create cities that functioned as spaces of capital accumulation. Cooke (1983: 145) hence concluded that 'planning performs its main functions by solving land use dilemmas . . . and smoothing the dynamics of land development'. The work of Scott (Scott 1988; Scott and Storper 1992) thus opened up another avenue of Marxist urbanism – one attempting to integrate planning and urban theory to analyse the role of planners in resolving conflicts between capital and labour by intervening in land markets (see P. Clarke 1989).

Marxist perspectives have consequently deployed to great effect in explorations of urban land use and segregation. Notably, theories of uneven development have provided the basis for exploring phenomena of 'block-busting' and gentrification, with the devaluation and subsequent redevelopment of specific urban tracts seen as a key means by which capital seeks the most profitable locations for its realisation (N. Smith 1979). But Marxist perspectives also provided a way of thinking through the evolution of the urban system (notably, the changing relations between cities) as well as exploring the importance of the state in the regulation of capital/labour conflicts at scales varying from the household to the urban region. In the context of this chapter, it is impossible to do justice to the rich diversity of Marxist urbanism. However, it is probably fair to suggest that all urban theory inspired by Marx rests on the idea that urban development can be understood only in relation to the 'bedrock' of capitalism. In short, capitalism is seen as the root of all urban problems, and provides the inevitable answer to the 'urban question'.

POSTSTRUCTURALISM, POSTMODERNISM AND 'OTHER' THEORIES OF THE CITY

Managerial and Marxist perspectives on city life are widely acknowledged as having rejuvenated and reawakened urban studies, and inspiring a fertile cross-disciplinary dialogue on city life. Yet such radical ideas were never subscribed to by all, and it is probably fair to say that the majority of urban researchers continued to work within positivistic traditions which prioritised agency over structure. Some of these urban researchers were openly critical of Marxism, suggesting that it was less of an explanatory theory and more of a political dogma. Yet some of the fiercest criticisms of Marxist urbanism were to emanate from those who sought to work with Marxist theories but ultimately found them too rigid to account for the range of differences and diversity characteristic of city life. For instance, while feminist critiques predated much of the work carried out in the name of Marxist urbanism, the radicalism of both encouraged a fecund dialogue between feminist and Marxist scholars. However, the diversity of feminist positions meant that while some feminist scholars felt gendered inequalities could be adequately explained within the plenary geography of capitalism, others were less convinced of the merits of a Marxist approach (see Walby 1997). In this sense, many feminist scholars concluded the experiences of women in the city are connected in important ways to their allotted place in the capitalist workforce (with the feminisation of particular economic sectors having important implications for women's mobility and visibility). Yet the primacy of capitalist work in shaping gender inequalities was questioned by others, who felt tales of class oppression deflected attention from crucial issues of sexism and gender domination in other social spheres (G. Pratt 1991; Valentine 1997). For some, the fact that gender inequalities seemed to be present in cities prior to the emergence of capitalism suggested that patriarchy (the system by which male values dominate female ones) needed to be considered as significant in its own right, albeit 'inextricably interwoven' with class, race and other axes of inequality (Bondi and Christie 2002: 293).

One implication – that both class and gender are significant in determining people's spatial freedom and mobility – indicates the limitations of Marxist perspectives which begin from the assumption that the social landscape of the city can be explained solely in terms of class. Yet another critique of Marxist urbanism emanated from arguing for postcolonial perspectives. Narrowly defined, postcolonial refers to the peoples and nations who have lived through processes of formal decolonisation, yet it has also come to denote a more wide-ranging understanding that we live in a world where knowledge forged in imperial times can no longer be regarded as appropriate (Sidaway 2000). In general terms, postcolonialism has emerged as a multidimensional critique of the hegemony of Western geographical imaginations, particularly those which divide the world into West and East and ascribe overarching characteristics to specific regions according to this duality. The work of Edward Said is often cited as a keystone of postcolonial studies given that it challenged mainstream Western imaginations of the Orient, and staked a major case for the construction of indigenous geographical knowledges. Yet the postcolonial critique is also relevant to other geographical scales, stressing that questions of race, ethnicity and power are important even in spaces which were never subject to colonial control. For example, it is possible to argue that cities like London, Paris and Berlin can be understood as postcolonial, in the sense that they are former imperial cities which are now home to varied diasporic communities and where relationships between different ethnic groups continue to be shaped by the ideologies and imaginations of Empire in one way or another (King 2004).

The relationship between the colonial and postcolonial is a complex one, emphasising the importance of exploring the specificity of the ideological forces shaping urban space. One important implication here is that the language and concepts used to describe certain Western cities (such as Chicago) cannot be regarded as appropriate for describing those cities beyond the West; it also implies that these may be inadequate for describing the diversity of cities within the West given their varied roles in geographies of imperialism and Empire. The idea that cities are more varied and diverse than the theories that geographers make is thus an important one, and suggests our understanding of cities needs to be informed by considering different cities in different contexts (Robinson 2002). This is clearly tied into ideas of urbanisation and globalisation which suggest that it is vital to pay attention to the flows and processes that shape cities throughout the world – not least in those non-Western cities which have tended to attract less interest within a scholarly literature more fixated on the cities of the West (Marston and Manning 2005; see also Chapter 5).

Feminism and postcolonial theories have thus become increasingly significant in human geography, as have other theoretical frameworks (e.g. queer theory, subaltern theory – see Hubbard et al. 2002) which emphasise questions of difference as they are played out in the identity politics of city life. In distinct ways, each rejects the certainties of Marxist structural and class-based analysis in favour of theories sensitive to other forms of difference. This openness to the world, and the idea that other forces bar capitalism are in play, is also a defining characteristic of post-structural thinking. Though difficult to define, post-structuralism's emphasis on questions of language, representation and power points to a different way of understanding the production of space, involving the entwining of immaterial and material forces (see Chapter 2). Notably, many of the key proponents of post-structural thought – Foucault, Derrida, Deleuze, Irigaray, Baudrillard – explicitly critiqued Marxist thought and sought to develop alternatives more flexible and open to the messiness of life. Foucault in particular developed a critique which destabilised the authority of the scholar, and posed important questions about the power of disciplinary (and disciplined) accounts of the social world. Critical of the totalising discourse characteristic of social science,

Foucault argued for the recovery of subjugated knowledges ('those that have been disqualified as inadequate to their task or insufficiently elaborated: naïve knowledges, located low down on the hierarchy' – Foucault 1980: 82). In the wake of such Foucauldian critique, it has been difficult for researchers to argue that they have a privileged gaze, or to essentialise difference in the name of generalisation.

Widely cited in urban studies, the rise of post-structural thought has (directly or otherwise) encouraged urbanists to develop accounts of city life somewhat at odds with Marxist ones (and, for that matter, the neat models of urban ecology). Chief here has been a concern to move beyond totalising theories and embracing the richness of the local and the particular. By way of example, we might briefly consider the issue of gentrification – the phenomena of working-class populations being displaced from inner city districts as housing areas are appropriated and redeveloped by the more affluent. Marxist perspectives, and most notably the rent gap thesis proposed by Neil Smith (1979), indicated that gentrification represented the movement of capital back to the city, suggesting that the gap between actual ground rent and potential ground rent in such areas provided the incentive for capital investment in inner city areas. In turn, this could be seen to be related to phases of economic growth and decline, with investors engaging in property development when other areas of the economy appeared sluggish. In contradistinction, other commentators considered the increased *demand* for inner city living among diverse social factions as the key driver of gentrification. In the work of David Ley (1996), for instance, it is the aesthetic disposition of particular actors – especially artists – that is described as encouraging this valorisation of 'mundane' and run-down areas. Though lacking economic wealth, this group is regarded as rich in what Bourdieu terms 'cultural capital', prompting other groups to cluster around these artistic communities. For Ley (2003), this points to the intersection of economy and culture, and the limitations of explanations which seek monocausal explantion.

Hence, the pre-professional, creative middle class have been identified as especially significant in processes of gentrification (van Wessep 1994; Ley 2003). Yet there are other factions who have been identified as important, not least female-headed households and single women in paid employment (Bondi 1991; Mills 1993). In many instances, gentrifying households are dual income couples who have remained childless for personal or career reasons, with gay and lesbian groups often depicted as instrumental in creating geographies of gentrification (Lauria and Knopp 1985). In each case, it has been suggested that these sections of society have cultural interests and housing demands which can be satisfied by city centre living. For example, many gentrifiers work in business or creative industries in the central city and have long or irregular hours and want to live close to work and the cultural and entertainment facilities offered by the city centre (Hamnett 1991). In part, the attraction of living in the inner city for such groups is also that they feel part of the creative life and 'buzz' of city life, and are participants in putative 'urban renaissance' (see Chapter 6). For instance, Butler's (2003) research suggested that a principal attraction of living in some central neighbourhoods in London is their 'metropolitan habitus': the cosmopolitan and cultured outlook that unites gentrifiers in an 'imagined community' of like-minded individuals. Distinctions of Self/Other may also be worked through in the midst of such processes, encouraging those who can afford to do so to distance themselves from populations which they regard as threatening because of their apparent difference (Sibley 1995).

In such ways, the gentrification literature suggests that gentrifiers are attracted to the central city for a number of reasons, and that although economic structures are important, varied questions of lifestyle and identity may be equally or even more significant. This type of conclusion demonstrates the limitations of Marxist perspectives which tended to view social and cultural processes as mere side-effects of capitalism. Caricaturing the work of Marxist theorists (such as Harvey), Barnes (2005: 67) argues that culture was frequently reduced to an epiphenomenon, 'sloughed off' as not essential to understand the 'real' business of the urban economy. Yet a widespread attempt to rethink the relationship between economy and culture, and to recognise that the lines between culture and economy are not so sharply drawn, impelled many dyed-in-the-wool Marxist geographers to reconsider the importance of cultural values and practices as constituents of political and economic life. This realisation, when coupled with the insights of post-structural philosophy, was to be a major factor encouraging urbanists to develop accounts of the city more attuned to cultural difference and diversity (see also Chapter 2 on the 'cultural turn').

Yet there was a perhaps more significant reason that many urbanists began to turn away from class-fixated Marxist theories. Simply stated, by the late 1970s the industrial city had mutated into a very different species, and theories evolved in the context of industrial cities no longer appeared to hold. One significant observation here is that many Western cities were decentring. New office complexes, science parks, retail malls and leisure parks were springing up around the periphery of most cities, while a rash of gated communities (privatopias) produced new residential foci. In some instances, this peripheral urban development created new urban centres or cores which effectively become *edge cities* (a term originally coined by Joel Garreau 1991). This decanting of residential, leisure and workspace to the periphery of Western cities has been a widely noted phenomenon, and while this tendency has been tempered in much of Europe by planning controls, peripheral development has taken extreme forms in the United States, often accompanied by wholesale disinvestment in the central city. This leads to the so-called 'doughnut syndrome', where the majority of the city's job creation and consumer spending is located not in the centre, but at the edge: the city literally turns 'inside out'. In the metropolitan region of Dallas/Fort Worth, for example, the core of the region accounts for only 10 per cent of the jobs: the majority are in new growth corridors and urban 'centres' dotted along interstate highways (W.A.V. Clarke 2002).

For such reasons, the post-industrial city has often been described as a *centreless* urban form. This decentring appears to be connected to important changes in the economy of cities, with the economies of scale which were important in the context of 'Fordist' mass production giving way to 'post-Fordist' economies of scope. In simple terms, this meant that large, inflexible businesses sought to retain their profitability by becoming leaner and more flexible. In part, this was achieved by firms outsourcing certain stages of production through subcontracting. In practice, this meant that firms and businesses were seeking to move out of older industrial districts to more specialised 'flexible production districts' where firms benefited from proximity to subcontractors and firms in related industries (A. Scott 1988). Given these districts could be quite specialised (as in the crafts and furniture production that predominated in the Third Italy region), there was little requirement for these districts to remain in accessible urban cores: on the contrary, many of these districts decentred, locating adjacent to the neighbourhoods from where most of their knowledge-rich workers originated.

But while many commentators have emphasised that the post-industrial city is characterised by new types of industrial space, what is perhaps most significant about post-industrial cities is that they are organised around consumption rather than production (Zukin 1998). Manifest in a plethora of spaces of mainstream and alternative consumption (malls, multiplexes, cafes, festival marketplaces, nightclubs, super-casinos, heritage parks, museums), the implication is that the post-industrial city is subject to a new logic of social control in which individuals were divided into consumers (the 'seduced') and non-consumers (the 'repressed') rather than workers and the unemployed (D.B. Clarke 2003). Herein, consumers' need for commodities has seemingly been replaced by desire: a desire that cannot be sated, only fuelled. According to Bauman (2001), consumer *wants* have been wrenched out of the grip of *needs*, with seduction and temptation stimulating capricious and conspicuous consumption. Yet this is a process replete with contradiction: in an era of rampant job insecurity and global risk, consumption promises security, but fuels that very insecurity:

> Seeking security through consumer choices is itself a prolific and inexhaustible source of insecurity. Finding one's way amidst the deafening cacophony of peddlers' voices and the blinding medley of wares that confuse and defy sober reflection is a mind-boggling and nerve-wracking task. It is all too easy to be lost, even easier to make costly mistakes . . . I have no way to say whether the blouses hawked and huckstered this summer are smart and flattering or downright ridiculous, uglier than the blouses in the next shop. But . . . this is not really my business – I am not to be blamed for wearing one; there is a designer/brand label on my neck or the designer/ brand logo on my chest for everyone to see and shut up . . . Designers breathe supreme and absolute authority. There is no point in contesting that authority, but a lot of sense in hiding behind it. And there is the 'return to shop if not fully satisfied' promise, the 'money back' guarantee.
>
> (Bauman, cited in Rojek 2005: 303)

We thus allow ourselves to be seduced into the consumer fantasy that we can buy security and freedom, irrespective of the fact that there are always new brands, commodities and ideas to buy. The design and appearance of consumer spaces belies this contradictory logic, with enchanting and seductive architectures encouraging a playful form of consumerism that promises security and satisfaction, but leaves our desire unsated. Significantly, consumer spaces often trade on the notion of individuality and the myth that the desires of different groups can be accommodated via practices of *mass* consumption. As such, consumer architectures often celebrate minority cultures, and may be themed around specific ethnicised identities (e.g. Banglatown and Chinatown in London). Yet at the same time, surveillance is ever present in these settings, excluding troublesome non-consumers and the credit-poor from the leisured consumption spaces which pockmark the post-industrial city and act as its chief loci of economic growth. For the excluded, the city appears an ever-more repressive and violent space, offering fewer and fewer public spaces of democratic social interaction (M. Davis 1990). Indeed, it has been widely noted that the presence of non-consumers in many public spaces triggers panic and fuels Zero Tolerance policing, with the homeless, prostitutes, drug-dealers, itinerant traders and youths all having been subject to policing which circumscribes their spatial freedom (Sibley 2001) (see also Chapter 3).

In sum, the post-industrial city is regarded as a more flexible, complex and divided city than its predecessor, with the ordered and production-based logic of the industrial era giving way to a more invidious mode of social control based on one's role as consumer-citizen. The result is a patchwork city of different ethnic enclaves, consumer niches and taste communities, spun out across a decentred landscape where the boundaries between city and country are hard to discern. Given this, a rash of new terms emerged to describe the post-industrial city and its attendant spatial forms: the splintered city, the edgeless city, the urban galaxy, the spread city and so on (Taylor and Lang 2004). But above all else, this form of city became identified as *postmodern* (see Box 1.3).

Taking on board a range of ideas about the triumph of consumerism over production in Western societies, theories of the postmodern city imply that it not only looks different from its predecessors, but also it works according to different 'logics'. This has impelled several Marxist urbanists (most notably Harvey and Soja) to revisit their ideas of urban political economy. For example, Harvey (1989b) takes Los Angeles to be sympto-matic of contemporary urbanism as it is a city in which there is a plethora of cultural signs and images which come together to form a melange within which there seems to be no overall order. However, Harvey suggests that this is largely an outcome of a new form of capitalism which he has termed 'flexible capitalism'. Harvey makes this claim on the basis that he can discern

Box 1.3 THE POSTMODERN CITY

Although postmodernism is a term used in a number of diffuse and complex ways, it is principally used by urban geographers to describe the type of Western city emerging in the late twentieth century in response to a complex range of economic, social and political restructurings. Soja (1989) gave a flavour of these changing urban geographies in his book *Postmodern Geographies*, which presented Los Angeles as the quintessential postmodern city. His description of the diversity of downtown LA gave some initial indications as to the new forms of demarcation emerging in the city:

> There is a dazzling array of sites in this compartmentalised corona of the inner city: the Vietnamese shops and Hong Kong housing of a redeveloping Chinatown; the Big Tokyo financed modernisation; the induced pseudo-SoHo of artist's lofts and galleries . . . the strangely anachronistic wholesale markets . . . the capital of urban homelessness in the Skid Row district; the enormous muralled barrio stretching eastwards toward East Los Angeles . . . the intentionally yuppifying South Park redevelopment zone hard by the slightly seedy Convention Center, the revenue-milked towers and fortresses of Bunker Hill.
>
> (Soja 1989: 239–240)

What Soja described was a new type of city permeated by divisions and fractures which could not even have been guessed at by the members of the Chicago School, let alone in Marx and Engels' day. Soja's focus on LA is significant as it has become the 'ur-city' of postmodern urban theory, spoken of variously as 'a prototype of our urban future' (Dear 2002: 10), 'a polyglot, polycentric, polycultural city' (Dear and Flusty 1998: 52) and 'one of the most dramatic and concentrated expressions of the perplexing theoretical and practical urban issues that have arisen at the end of the twentieth century' (Scott and Soja 1996: viii) (cited in Brenner

1999). LA has thus served as the basis of a plethora of models of the postmodern city, with a putative LA School emerging among those urbanists and geographers who have been most influential in developing theories of postmodernity. What is particularly interesting about the LA School is their almost uniform pessimism and identification of LA as teetering on the verge of dystopian meltdown. In particular, Mike Davis (1990) has emerged as a widely heralded spokesman on the death of cities, his ideas of urban life being undermined by ecological disaster, terrorism, inequality and dysfunction informed by his own reading of Los Angeles' dystopian landscapes.

Key reference: Soja (1989)

close similarities in the way the postmodern city is produced and the way capitalism works: 'I see no difference in principle between the vast range of speculative and . . . unpredictable activities undertaken by entrepreneurs . . . and the equally speculative development of cultural, political, legal and ideological values' (Harvey 1989b: 344). Related to this, he has argued that much of the postmodern city appears to be produced by and for institutions which are clearly capitalist in nature (multinational architectural firms, property-developers and financial institutions).

Rather than being an imposition of different cultural values and desires on the city (a postmodernism of resistance), Harvey depicts changes in the urban landscape as driven by big businesses which manipulate tastes to create profitable urban lanscapes (a reactionary postmodernism). As Cloke et al. (1991: 182) note, Harvey suggests that the practices of 'pastiche' employed in postmodern art and architecture might be regarded as 'nothing more than an extreme manifestation of the relentless and structurally determined quest for new and unusual commodities to sell'. In an era in which sites of consumption are increasingly rationalised settings, it is argued that the postmodern theming of places is a means of 're-enchantment' – replacing the impersonality and instrumentalism of consumption (Ritzer 1999). The spectacular form of many postmodern enclaves and spaces is thus significant, with Harvey suggesting that postmodern design acts to cover over some of the problematic practices of capitalism (see Plate 1.2).

Plate 1.2 Postmodern urban forms: a complex, playful architecture of surface and seduction (La Defense, Paris) (photo: author)

Harvey also argues that the postmodern city is designed to mystify people by making things appear more exciting, more individual, more open, more human scaled and yet still house corporate, multinational firms which carry on their practices as before.

Harvey's distinctive interpretation of the postmodern city was enthusiastically greeted in many quarters, and, read alongside Soja's (1989) *Postmodern Geographies* and Jameson's (1984) essays on postmodern art, suggested an important link between the logics of capitalism and the emergence of a new type of city. Yet this interpretation was also subject to critique, attacked for being insufficiently postmodern in approach. Dear (2000: 76), for example, claims that both Harvey and Soja established 'profoundly modernist' accounts of the postmodern city, characterised by a lack of attention to 'the consequences of difference' (Dear 2000: 766). Among the neglected differences identified by Dear are those of 'gender and feminism' which he claims are almost completely absent within Harvey's (1989b) *The Condition of Postmodernity*. Concurring, Massey (1991b) suggests that when Harvey (1989b) describes the experience of being in the postmodern city as having no stable bearings and being unable to impose order on one's surroundings, this is very much a white, male, middle-class academic view. This failure to account for and see the world through Other eyes means that Harvey's Marxist version of postmodernity constantly falls back into the language and logic of modernism, and imposes an order on the postmodern city which simply does not exist (Watson and Gibson 1995). Here, it is important to note that one of the key precepts of post-structuralism is that being open to difference, and adapting different modes of representing the world, can lead people to adopt very different behaviours and instigate a 'politics of change'.

In the eyes of many, this apparent lack of 'critical self-reflection about the author's own epistemological stances' (Dear 2000: 76) has devalued Harvey's highly nuanced interpretation of postmodern cities. Michael Dear (2000: 78) complains, for example, that while 'taking an axe to postmodernism, Harvey leaves his own historical materialism almost totally unexamined' and offers a starkly modernist, structuralist analysis of postmodernity which seeks to 'get behind' the fragmented features of postmodernity to identify its 'essential meanings' (Harvey 1989b: 74). Barnes (2005) likewise suggests that Harvey employs an interpretation of the relationship between culture and economy which appears quite close to a classical 'structuralist' one, whereby culture is seen to be determined by economic processes. Even in later work where Harvey has sought to respond to his critics by exploring questions of difference in a more sustained manner (e.g. Justice, Nature and the Geography of Difference 1996; Spaces of Hope 2000), Barnes (2005) suggests his account remains rooted in Marxism and fails to develop a more nuanced account of culture in the making of contemporary cities. Somewhat similarly, Soja's (1996) Thirdspace presented an extended engagement with feminist, postcolonial and subaltern theory, yet remained open to accusations of tokenism given the unshaken belief that the changing forms of the city could be explained away with reference to transitions in the nature of capitalism (see Latham 2004).

What appears to be at stake here is whether the contemporary (postmodern) city can be usefully explained with reference to the theories evolved in the context of the modern city (as Harvey and Soja's work implies). Here, we need to remain mindful that some of the characteristics taken to define the postmodern city were actually those of the modern city. Underlining this, Savage et al. (2003) suggest that movement, restlessness and dynamic transformation were essential qualities of the modern urban experience as much as the postmodern (cf. Wirth 1938; Harvey 1989b). Moreover, one has to be mindful that many postmodern theories have been worked through in the context of Los Angeles, a city whose ethnic diversity and residential segregation is 'almost unique' (see Johnston et al. 2006). In this regard, some commentators remain sceptical that new theories are needed to explain the spatial forms of the postmodern city, for example see W.A.V. Clark's (2002) empirically derived bid-rent curves describing the evolution of a decentred, polycentric metropolis.

Notwithstanding this, much appears to have changed about the city since the mid 1970s, with cities having undergone dramatic transformations in their physical appearance, economy, social composition, governance, topographical shape and cultural vernacular (Savage et al. 2003: 32). Taking this into account, many commentators insist that we need new theories more in-tune with contemporary urbanism, and that 'modern' theories are simply past their sell-by date. For instance, Dear and Flusty (1998) attempt to develop a new theory of postmodern urbanism which acknowledges what is really happening in contemporary cities where:

> Urban process is driven by a global restructuring that is permeated and balkanized by a series of interdictory networks; whose populations are socially and culturally heterogeneous, but politically and economically polarised; whose residents are educated and persuaded to the consumption of dreamscape even as the poorest are consigned to carceral cities; whose built environment, reflective of these processes, consists of edge cities, privatopias, and the like, and whose natural environment, also reflective of these processes, is being erased to the point of unlivability.
>
> (Dear and Flusty 1998: 59–60)

This leads them to offer a 'model' of the postmodern city which is very much at odds with the centred and ordered models of urban ecology (see Figure 1.1). Herein, 'keno-capitalist' processes create a pseudo-random patchwork of balkanised industrial, consumer and residential landscapes.

But even in the case of Dear and Flusty's writing on urban form, we can detect an attempt to theorise and explain the city in terms of existing and established theories (notably, Marxism). No matter how much Dear and Flusty (1998) appear to embrace postmodernism's suspicion of

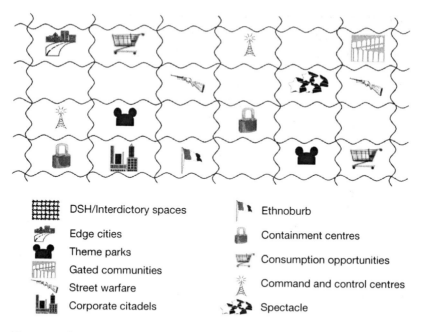

▦ DSH/Interdictory spaces	▐▌	Ethnoburb
Edge cities	🔒	Containment centres
Theme parks	🛒	Consumption opportunities
Gated communities		
Street warfare	📡	Command and control centres
Corporate citadels	✯	Spectacle

Figure 1.1 The postmodern urban landscape (after Dear and Flusty 1998)

metanarrative, they quickly fall back into the language and assumptions that underpinned explanation of the modernist city, remaining ideologically wedded to a Marxist narrative in which historical materialism provides the bedrock of theoretical explanation. Following Lake (2005: 267), I take such contradictions as evidence of a theoretical crisis in which urban scholars have sought to move beyond the twin horrors of positivist quantification and Marxist abstraction, yet fear the 'capriciousness' of post-structural, postmodern and post-positivist approaches which insist on the particularity of each and every event. Hence, while there is a widely noted dissatisfaction with the 'will to abstraction' which forced the city to conform to abstract models, categorisations and languages, urban scholars have often fallen back on these very forms of abstraction in their attempt to comprehend new forms of urbanity. Lake (2005: 267) accordingly concludes that in the last quarter of the twentieth century, urban geography reached a theoretical nadir, with the urban becoming little more than 'an epiphenomenon, a barely significant by-product of structural forces'.

CONCLUSION

In this chapter, we have taken a somewhat breathless tour through two centuries of thinking about the city, alighting on some of the most celebrated and important urban theories. As we have seen, these theories have been tailored to answer particular questions about cities, such as what they do, how they work and why they exist. In essence, however, all concern the urban question – that is to say, they seek to say something about the relationship between the city and society. As was argued at the outset, many of these theories have borrowed ideas and languages derived from other disciplines – for example, biology, mathematics, economics, political science and so on. This has meant that urban geographers now have a rich lexicon of terms and concepts at their disposal. Yet for all this, it is difficult to discern any *progression* of urban theory. The debate around gentrification provides an obvious illustration: while alternative explanations have been proposed for this phenomenon over time, there remains little agreement about the respective importance of culture or economy in this process, which is regarded variously as a movement of people back to the city or a movement of capital back to the city. Similarly, while the identification of a putative postmodern city has encouraged the development of postmodern urban theories, these have not totally supplanted modern ones, which continue to have proponents and supporters. Circling around the same issues to increasingly little effect, we have seen that many theories of the postmodern city remain ideologically wedded to a political economy that seems unable to offer any new insights into the nature of urban life.

Commenting on this lack of progress in urban studies, Nigel Thrift (1993: 228) contended that urban studies had reached something of an impasse by the 1990s, being characterised by 'recycled critiques, endlessly circulating the same messages about modernity and postmodernity'. In a review of theorisations of the city, Michael Storper (1997) echoed this, arguing that while urban geography agrees on many of the central aspects of contemporary urbanisation (e.g. the importance of consumerism, the role of global flows of capital, knowledge and goods, and the flexibilisation of production), no theory seems able to put these phenomena together in any meaningful way.

In addition to the failure of urban theory to 'speak to' the urban conditions that characterised the city of the late twentieth century, by the 1990s it was apparent that what passed for urban theory was increasingly devoid of urban specificity. Indeed, it is possible to argue the geographies so eloquently described and explained by members of the LA School were not urban geographies *per se*, but more broad-brushed accounts of the way space was restructuring in accordance with the dictates of postmodern times. Hence, in contrast to the many rural geographers and sociologists who stressed the continuing need for new theories of rurality in the face of rapidly changing socio-spatial relations (see P. Jackson 2005), urban geographers seemed more concerned with contributing to more general theories of spatiality. Others, as Wyly (1993) notes, were simply content to get on with 'doing' urban studies, examining issues, populations and problems that just happened to be located in urban areas, rather than thinking through what was distinctively urban about the nature of the research they were undertaking.

As I detailed in the introduction to this volume, there were accordingly many reasons to be pessimistic about the future of urban geography at the dawn of the twenty-first century, with fewer and fewer geographers appearing interested in making any substantive contribution to urban theory. Yet at the same time, human geography remains a vital and vibrant sub-discipline, characterised by new and exciting ideas about the nature of space and place. Consequently, I would argue that – even if contemporary urban scholarship seems unwilling or unable to say anything in depth or profoundity about the spatiality of cities – there are many concepts and ideas in contemporary human geography which are highly relevant to developing new understandings of the 'city' as a distinctive and important spatial formation. Developing this argument, the remainder of the book will seek to review four recent approaches in geography which have significant implications for how geographers study the urban: a 'representational turn'; a (renewed) interest in embodiment and performance; the emergence of a 'post-human' geography, and an interest in the formation of a network society. In different ways, each of these provides a distinctive take on the city's distinctive and immanent materiality: taken together, they arguably provide the basis for a rejuvenated urban geography.

FURTHER READING

As has been stressed throughout, this book is about urban theories and conceptualisations of the city. For those who are studying urban geography for the first time, it is recommended that this book is read alongside an introduction to urban geography. Many of these are organised thematically, and deal with economic, social and political aspects of city life: John Short's (1996) The Urban Order is particularly recommended, while Tim Hall's (2001) Urban Geography provides a succinct and balanced overview. Paul Knox and Steve Pinch's (2001) Urban Social Geography focuses (as its title implies) on social segregation and consumption issues, but contains useful material on urban policy and politics. The Open University series Understanding Cities comprises three books, the first of which is concerned with urban definitions, the second with urban order and the third with urban movement. The latter volumes are referred to later in this book, but in this context it is worth tracking down the first volume (Massey et al. 1999) as an introduction to debates concerning the definition of the city. For those interested in discussions of the origins of urbanism and changing city forms, Harold Carter's (1988) An Introduction to Urban Historical Geography remains a useful overview and Spiro Kostof's (2001b) The City Shaped and (2001a) The City Assembled provide a lavishly illustrated history of city design.

In relation to the material discussed in this chapter, Savage et al. (2003) Urban Sociology, Capitalism and Modernity provide a good interdisciplinary account of modern and postmodern theories. Extracts from the writings of key theorists can be found in LeGates and Stout (eds) (1996) The City Reader. Finally, it is well worth exploring the writing of two of the figures whose work has exercised such an influence on the recent trajectory of urban studies – Manuel Castells and David Harvey. Andrew Merrifield (2002) profiles each of these in his excellent Metromarxism; Ida Susser (2002) has compiled a useful Castells Reader while Jones et al. (2005) provide a critical overview of Harvey's work.

2

THE REPRESENTED CITY

The city is a discourse and this discourse is truly a language.

(Barthes 1975: 92)

In the late 1980s and 1990s, there was much talk of a 'cultural turn' in human geography. Though subsequent evaluation suggests there was not one turn, but many (see Philo 2000), the key notion underpinning the cultural turn – that culture needed to be taken seriously – was one that was widely embraced by geographers. In particular, geographers studying the built environment began to acknowledge the need for an interpretative approach that would disclose the intersubjective meanings and symbolism of the urban landscape. Indeed, the identification of urban landscapes as legitimate objects of study was crucial to moves within the 'new' cultural geography intended to shake off the rural and historical predelictions inherited from the Berkeley School of cultural geography in the United States (P. Jackson 1989; Crang 1998).

However, the cultural turn had another key impact on the trajectory of urban geography – namely, that it encouraged exploration of the image of the city and its representation in a variety of media. Crucial here was the suggestion that space is constructed both in the realms of discourse and practice, and that it is impossible to conceive of any space outside the realms of language. Indeed, Jones and Natter (1997) suggest it is only through representation – words, images and data – that space exists, with all spaces being both 'written' and 'read'. A widespread approach to urban geography since the late 1980s has thus been to conceptualise the city as a text (Duncan 1990). Simply put, this textual metaphor has encouraged researchers to explore the 'discourses, symbols, metaphors and fantasies' through which we ascribe meanings to the city (Donald 1999: 6). This approach takes inspiration from a wide range of sources, including the literary theories of Roland Barthes, the cultural materialism of John Berger and Raymond Williams, Michel Foucault's ruminations on the power of discourse and Jacques Derrida's philosophy of deconstruction.

In this chapter, I thus want to explore some of the key facets of this 'representational turn' in human geography, describing how it has changed the ways that many urban geographers look at, visualise and generally make sense of the city. This overview considers geographers' attempts to unpack the images of cities conveyed in fiction, film and other media, as well as the efforts made to situate urban mythologies within a richer theoretical context. Moving from these efforts to 'read' cities, I then consider how the incessant production of new urban images and stories constantly revises our perception of particular cities via processes of place marketing and identity-making. I begin, however, by exploring some of the (long-established) knowledges of the city that frame these attempts to rewrite the city.

URBAN MYTHS AND IMAGINATIONS

The idea that cities are symbolicly associated with particular values, lifestyles and ideas is a long-standing one. In contrast with the countryside, which is principally known through the myth of the 'rural idyll', dominant under-standings of cities tend to be framed within two prominent myths. The first of these is an anti-city myth which depicts cities as a nadir of human civility. In this anti-urban myth, the city is associated with sin and immorality, with a movement away from 'traditional' order and mutual values. Biblical references to Sodom and Gomorrah, destroyed by God for sins against His law, demonstrate the antiquity of this imagination of the urban, while more recently the sheer ugliness of Victorian industrial cities, and the 'urban decay' experienced within contemporary urban areas have inspired many journalists, writers and artists to catalogue a range of urban ills. John Ruskin, for example, the nineteenth-century essayist and anti-industrialist, said that cities are 'loathsome centres of fornication and covetousness — the smoke of sin going up into the face of heaven like the furnace of Sodom' (Ruskin 1880, quoted in Short 1991: 45). Here, the anonymity which the city often provides, instead of providing an opportunity to be whatever one wishes to be, becomes associated with a sense of alienation, of not belonging.

In many ways, this myth is backed up with reference to many of the ideas of urban anomie and isolation developed by urban sociologists in the late nineteenth and early twentieth century (see Chapter 1). Yet it is also a myth bolstered by the persistence of the myth of the rural idyll, in which the city is counterposed with a timeless, harmonious picturesque rurality characterised by an absence of social problems and the fostering of good physical, spiritual and moral health (Holloway and Hubbard 2001). The redemptive qualities of rural living accordingly informed the emergence of a town and country planning movement which attempted to reintroduce some of the characteristic features of country living into the city: Ebenezer Howard's Garden City ideal, for instance, argued for small, self-contained communities set against a green background, while even Le Corbusier was concerned to reintroduce nature to cities (Hall 1988). Contemporary experiments in 'neo-urbanism', such as Poundbury in Dorset or Seaside in Florida, also romanticise many aspects of small-town and rural living, suggesting these produce cohesive and friendly communities (McCann 1995).

The idea that rural living is therapeutic and morally uplifting therefore exists in opposition to mythologies which emphasise the city's role as a locus of immorality, criminality and disorder – 'a temptation, trap and punishment all in one' (Nochlin 1971: 151). Such views may be supported with reference to the onslaught of media stories which focus on 'urban problems' of gun crime, drug-dealing and antisocial behaviour. This type of mythology is evident in any manner of songs, computer games, music videos, books, films and comic strips which script the city as a combat zone in which different clans or tribes face a daily battle for survival: gangsta rap, for example, has frequently been condemned for its glorification of gang violence on the streets of South Central LA, Compton and Detroit, yet is interpreted by many to be an authentic account of life in the post-industrial city. Other representations are perhaps not meant to be taken too seriously, yet similarly reinforce ideas that cities are sites of disorder. For example, Pierre Morel's (2003) *Banlieue 13*, which is set in a Parisian suburb 'sometime in the near future', imaginatively locates gang conflict in a dystopian urban landscape on the periphery of Paris. The fact that the best-selling computer game of all time, *Grand Theft Auto: San Fernando*, dwells lasciviously on the terrors of urban life is no mere coincidence either, given it taps into long-standing fantasies and fears of urban disorder.

Developing such arguments, Short (1996) concludes that the *cinematic city* is often a site of anonymity, crime and vice (Box 2.1). For example, in *film noir* (literally 'dark film') the city becomes a brooding, threatening presence, a place of isolation and fear. This often takes exaggerated form in films where the city's lawlessness can be contained only by crime-fighting superheroes, such as *Batman's* Gotham City or *Spiderman* in New York. Science fiction representations of the future city (e.g. *Things to Come, Alphaville, Nineteen Eighty-Four* and *Blade Runner*) similarly paint a dystopian image of urban life (see Kitchin and Kneale 2001), contrasted with the simple and idyllic nature of rural life (e.g. as exhibited in Merchant Ivory costume dramas). This anti-urban/pro-rural representation is very common in cinema, far outweighing the small number of rural horror films as *The Straw Dogs, The Blair Witch Project* or *The Texas Chain Saw Massacre*.

Box 2.1 THE CINEMATIC CITY

The connections between cinema and the city are many and various. For instance, it has been suggested that the synergy of film and the city was cemented in cinema's earliest years, when film was used as a medium which allowed urban citizens to make sense of the city: in short, film captured the restlessness and frenetic pace of the city in ways that other media could not. Likewise, cinema-going and (latterly) the consumption of video, TV and DVDs have been thoroughly integrated in the rituals of city life (Barber 2002; Hubbard 2002). Spaces of cinema – from the earliest spaces of film exhibition to the digital cinemas and multiplexes of the contemporary era – consequently tell us much about the changing spatial forms and practices of urban life (see Plate 2.1). But, in representational terms, what is especially interesting about the relationship between film and the city is that we have come to experience the city cinematically. In effect, this is to suggest that films have changed not just the way we *look* at the city, but also the way we *act* in cities. In part, this is a consequence of films' mobile gaze, which lends the medium its veracity and believability, mirroring our own experience of urban spaces (Friedberg 2002). In another sense, it is because film-makers work in traditions of narrative cinema which impel us to identify with those whom we see on the screen, and to see the city as they do. Taken together, it is clear that film has had an extremely important influence on our understanding and conceptualisation of cities. By way of example, one can trace the influence of specific films on architects and planners, who have often designed cities as if they were 'stage sets' waiting to be populated by a cast of thousands. For instance, it is surely notable that many of those responsible for the post-war planning of British cities adopted a sleek, modern aesthetic which had first come to prominence in science-fiction representations of the future city. In a rather different respect, it is interesting to reflect on the influence of *film noir* on urban nightlife. Krutnik (1991: 23) explains that the *noir* city of Hollywood's thrillers of the 1940s and 1950s 'is a shadow realm of crime and dislocation in which benighted individuals do battle with implacable threats and temptations' (a theme that continues in modern films such as *Sin City*). Playing on our own fears and anxieties about Others and Otherness, *film noir* locates social threats in the shadows and fog of an unknowable and unmappable city. *Noir* themes of unknown people preying on others (and especially women) in the dark spaces of the unsafe city have arguably been important in implanting fear of the city at night among certain populations, and perpetuating the idea that cities are spaces of isolation and anomie (Slater 2002; Farish 2005).

Further reading: D.B. Clarke (1997)

In sum, the persistence of anti-urban myths means that the countryside is frequently imagined to be a more pleasant and idyllic place to live than the city. In the West, such views have encouraged processes of counter-urbanisation and selective out-migration as those who can afford to buy into the rural idyll leave the city for the countryside. More recently, however,

Plate 2.1 Star City multiplex, Birmingham: the multiplex cinema is bound into rituals of urban life in a variety of ways, both mirror and mould of urban society (photo: author)

it is apparent that many populations are returning to the central city, drawn by the combination of competitive property prices and talk of the urban renaissance. As we saw in Chapter 1, the idea of urban renaissance relies upon the valorisation of certain pro-urban myths, not least the notion that cities are attractive, vibrant and cultured places to live. In this mythology, cities are deemed to offer multiple opportunities for business and pleasure, their multiple social worlds perpetually throwing up new ideas, food, music and fashion. For the hip and trendy, cities are therefore the place to be. The idea that cities are somehow unpredictable, unknowable and even dangerous may furthermore be an attraction for such returning populations, not least those young, creative individuals who imagine themselves as pioneers on the 'urban frontier' (N. Smith 1996).

Accordingly, it is possible to identify both pro-urban and anti-urban myths, each of which takes form in relation to, or in opposition to, the other (Lees and Demeritt 1998). If the city is imagined through antithetical notions of desire and disgust, it is perhaps useful to examine the basis of these myths. These are not solely recent inventions, having developed over hundreds of years. For instance, in the classical Greek and Roman

periods, urbanism was associated with the idea of 'civilisation', planting
the seed of the myth that the city is the seat of culture, learning, government
and civil order (Sennett 1994). Since that time, it has been an icon of progress, enlightenment and opportunity, a sentiment celebrated in the story
of Dick Whittington, for example. According to this tale, the eponymous
hero set out for London where, he had heard, the streets were 'paved
with gold'. He ended up becoming Mayor of London, of course, justifying
this view of the city as being full of opportunities for those bright or
lucky enough to be able to take them. In a more recent context, the UK
Conservative Party in the 1980s made great play of the entrepreneurship
and enterprise associated with the city, and implored the out-of-work to
'get on their bikes' to seek employment in the big cities. Similar myths
of urban opportunity abound in other nations, of course, not least in the
United States, where Hollywood's 'Dream Factory' reputation has inspired
many to seek out the bright lights of the big city in search of fame and
fortune. In the 1991 film Pretty Woman, the character of the street prostitute
played by Julia Roberts apparently shows this is a routine occurrence when
she is swept up off the streets by a wealthy businessman (Richard Gere).
While there are many subtexts in this film (e.g. heartless businessman
finds redemption in true love), the cliché of Los Angeles being a city where
you can make your dream come true is one that underlines the pernicious
myth of city as opportunity. Incidentally, another film with Julia Roberts
– Notting Hill (1999) – represents London as an urban village where people
from all walks of life mingle, allowing social and sexual relations to be
forged between unlikely protagonists.

In part, this romantic view of city life is supported by the idea that
cities are connected to what is happening elsewhere, opening a world of
opportunities. From their earliest origins, cities have frequently been
centres of international trade and migration, global media and international politics, a strategic role that is often imagined to be increasing
in an era of globalisation (see Chapter 5). Further, the city is viewed as
having dominion over the surrounding countryside. If cities are cultured
and vibrant, the farmed countryside has, as one side of its own mythical
existence, the stereotype of the ignorant and brutish yokel. This stereotype
contrasts sharply with that of the liberal-minded and educated towns-
man or woman, perpetuating the idea that rural folk are isolationist and
technophobic in contrast with the sharp-suited and quick-witted urban
dwellers who keep their 'finger on the pulse'. To underline this, Short

(1991: 43) paraphrases Marx in saying 'towns saved people from the idiocy of rural life'. While Marx was using the term idiocy to imply the isolationism of rural life, it is perhaps significant that many have taken his comment at face value, and reproduced the myth of the country 'bumpkin' or idiot.

Hence, the connection between urbanism and order, progress, power and learning is widespread, sustaining other pro-urban mythologies. Here we might think about the myth of the city as a cultural 'melting pot'. This valorises the very size of the city as providing opportunities for variety, social mixing and vibrant encounters between very different social groups. Because of this, the city may also be regarded as having a radical potential, where it is possible to challenge entrenched order and struggle for liberty and egalitarianism. The French and Russian revolutions, for example, had largely urban roots, while the student riots of 1968 were played out on the streets of major towns and cities, from Bologna to San Francisco. It is difficult, therefore, to specify a single pro-urban mythology if we consider that the city is associated with both order and revolution, where both of those things can be thought of as positive. Neither are these things solely urban. The countryside has historically been associated with the order imparted by feudalism (in Western Europe), while the contemporary countryside is becoming a place for protest against entrenched social order – in Britain, for example, fox hunting and experimentation with genetically modified crops have recently been challenged, while demands for access to privately owned moorland and mountain areas have been made for many decades.

In this light, we might conclude that cities are polysemous (i.e. signify different things simultaneously), to the extent that some of the attributes of anti-urban mythologies also appear in the pro-urban myth. The loss of 'traditional' values, for example, can signal either an irretrievable break- down of social order (as part of an anti-urban mythology) or a liberation from oppression (as part of a pro-urban mythology). These mythologies are complicated creatures, then, shifting as they are explored from vary- ing perspectives. By way of example, we might consider the film *My Son the Fanatic* (1999) directed by Udayan Prasad and written by acclaimed author Hanif Kureishi. This film follows Parvez, a Pakistani taxi driver (played by Om Puri), as he contends with the sudden conversion of his teenage son to Islamic fundamentalism, as well as his own dissatisfaction with life in the city of Bradford (northern England). Taking Parvez's point of view, the film explores his commitment to his wife and his developing love for a prostitute who eventually becomes a target of his son's religious and moral outrage. Based loosely on events that unfolded in Bradford in the mid-1990s, where (mainly) British Asian pickets sought to hound sex workers off the streets of Manningham (see Hubbard 1997), the film progresses across different urban landscapes (e.g. the café where Parvez meets his taxi driver colleagues, the local mosque where Parvez's son falls in with a charismatic convert to the fundamentalist cause, and the industrial wastelands in which sex is transacted for money) as it explores the imbrication of urban and personal identities. Bradford therefore becomes a key actor in this unfolding narrative, and emerges as a hybrid space whose identity is highly ambiguous. Shown to be a space of conflict, violence and endemic racism, Bradford is also shown as a city where interracial, cross-cultural friendships and alliances are creating new, positive understandings of what it means to be British.

While few films offer the social nuance and geographical texture of *My Son the Fanatic*, it is certainly true to say that most films cannot be neatly labelled as pro- or anti-urban. Underlining this, Pratt and San Juan (2004) suggest that *The Truman Show* and *The Matrix* simultaneously present and critique utopian visions of the city. In the former, the seemingly idyllic town of Seahaven (actually Seaside in Florida) is exposed as an inauthentic and intensely surveyed space which condemns Truman (Jim Carey) to a blissfully repetitive life; in the latter, Neo (Keanu Reeves) joins a group of unbelievers who choose to live in the 'real' world rather than the simulated utopia of corporate towers and shopping malls. In both, the city is seen to provide access to a multitude of seductive consumer goods and desires, but there is a price to pay: the loss of individuality. Through a process of redoubling, these films emphasise both the positive attributes and significant downsides of contemporary urban life.

Such a blurring of urban mythologies of the city, making ambiguous the distinction between simply 'pro-' and 'anti-', perhaps better reflects the true complexity of the social experience and representation of urban places. Indeed, this is an argument that recurs in the literatures on geographies of sexuality. On the one hand, this literature conceptualises cities as offering an 'escape from the isolation of the countryside and the surveillance of small-town life' (Weston 1995: 274; see also Castells 1983; Lauria and Knopp 1985). In this way 'cities' have often attracted those whose sexual identities differ from the heteronormal (see Weston

1995; Bech 1998). Yet the safety of cities for gay populations has been repeatedly brought into question (Myslik 1996) and the rural has also been advanced as a site of sexual liberation (Valentine 1997; Phillips et al. 2000). Again, the relational constitution of town and country demands that we reject any straightforward interpretation of urban life as either empowering or repressive. The imaginative geography of urban and rural, while useful in framing discussions of urban geography, therefore has significant limitations that need to be carefully examined and worked through in specific case studies.

READING THE CITY

Given the existence of a wide range of (apparently contradictory) urban myths, urban geography has recently been replete with efforts to document the way that different urban imaginations are brought into being through different cultural forms and media. Here, geographers are building upon a long-standing tradition in the arts and humanities of subjecting novels, poems and drama to critical scrutiny. Written by those ostensibly skilled in the use and manipulation of language, these texts frequently offer deeply evocative accounts of life in particular times and places. Because of this, they offer rich pickings for geographers interested in examining the character of certain cities. Some authors have consequently been feted by historical geographers because of their ability to weave beguiling and vivid 'word-pictures' of life in the past. A notable example here might be the work of Charles Dickens, whose novels serve to map out the geographies of Victorian London in rich detail (see Donald 1999). Far from occurring in a blank landscape, Dickens' novels are played out in 'real' landscapes whose physical forms, topography and appearance are made legible through a thoroughly spatialised language. In many of Dickens' books, for example, the plot unfolds in a landscape that is thickly described as one of danger and dread, with the fog of the capital enshrouding malodorous characters.

Yet it is not only in the realms of urban history that literature may be a useful source for excavating urban meanings. Many contemporary novels offer topologically detailed accounts of life in specific towns or regions, and these may capture many facets of everyday life that are effaced in academic accounts based on survey or ethnographic work. This is perhaps most obvious in those novels where the setting is emphasised as a significant component of the storyline (rather than an inconsequential backcloth against which the story unfolds). For instance, Tom Wolfe's *A Man in Full* (1998), Bret Easton Ellis's *Less Than Zero* (1985) and Jay McInerney's *Brightness Falls* (1992) and *The Good Life* (2005) offer rich descriptions of the post-Fordist American metropolis, describing Atlanta, Los Angeles and New York respectively. Likewise, those seeking descriptions of life in contemporary London can dip into works by writers as diverse as Martin Amis, Zadie Smith, Hanif Kureishi, Nick Hornby, Geoff Nicholson, Michael Moorcock, Doris Lessing, Ian McEwan, John Lanchester or Helen Fielding. Such novels are valuable not only because they offer detailed descriptions of individual buildings, neighbourhoods and locales, but also because they locate particular social groups and individuals in these spaces, mapping out the fractures of social class, race, gender, age and sexuality which characterise city life (for interesting accounts of literary London, and other cities, see Simpson-Housley and Preston 1994).

Brosseau (1994) accordingly suggests that such literary sources are useful for mapping out the social geographies of specific cities. In many instances, these works claim credibility because the authors concerned are very familiar with the milieu of which they wrote. Yet to simply regard novels as a source of factual geographical information clearly ignores the way authors imbue the places they describe with imaginative characteristics. As Pocock (1981: 11) argues, the 'truth of fiction is a truth beyond mere facts' as 'fictive reality may contain more truth than everyday reality.' Underpinning this seemingly nonsensical statement is the idea that novelists and poets succeed in conveying and communicating the 'sense of place' that is immanent in given locations better than actually being in that location could. This idea relies on the fact that literature evokes the experience of being in place eloquently, with the intensely personal and deeply descriptive language used by the writer able to convey the elusive *genius loci* inherent in a place.

Creative writing has certainly enabled many authors to express the reasons why certain cities are special to them, or to convey the sense of loss they experience when the old city makes way for a new one. For instance, Baudelaire was proclaimed the 'lyrical poet of high modernism' by the cultural critic Walter Benjamin because of the ability of his allegorical verse to capture the ambivalence of living in Second Empire Paris. Most famously, his poem *La Cygne* (The Swan) mixes images of past and present Paris in a paean to the pre-Haussmann era, as this extract indicates:

Old Paris is no more (a town, alas,
Changes more quickly than man's heart may change);
Yet in my mind I still can see the booths;
The heaps of brick and rough-hewn capitals;
The grass; the stones all over-green with moss;
The *débris*, and the square-set heaps of tiles.
. . . Paris may change; my melancholy is fixed.
New palaces, and scaffoldings, and blocks,
And suburbs old, are symbols all to me
Whose memories are as heavy as a stone.
And so, before the Louvre, to vex my soul,
The image came of my majestic swan
With his mad gestures, foolish and sublime,
(Baudelaire, Les Fleurs du Mal, 1861)

Quoting Baudelaire allows us to convey something of what it was like to live in a city where (to paraphrase Marx) everything that was solid was 'melting into air'. Other high modernist novels – Alfred Döblin's *Berlin Alexanderplatz* (1929), James Joyce's *Ulysses* (1904) or John Dos Passos' *Manhattan Transfer* (1925) – convey a similar sense of ephemerality, fragmentation and change in the modernising city. Accordingly, the use of literary texts to reveal the richly subjective and ambivalent nature of urban space constitutes an important tradition within cultural geography (Donald 1999).

Yet Brosseau (1994) has insisted that the novel is more than just a resource from which we may glean geographical facts or richly subjective accounts of place. For him, it is crucial that geographers are 'receptive to what is different in the way novels write and generate particular geographies' (Brosseau 1994: 90). After all, works of fiction are never mimetic, in the sense that they innocently reflect the world as it is. Rather, they refract that reality according to the author's positionality within systems of literary production and consumption. As such, 'critical' cultural theories have also been deployed in studies of fictional city writing. These suggest that texts are implicated in the reproduction of society, positing a recursive and complex dialectic between human agency and structure mediated through texts. This materialist conception of culture was one that was developed by the forerunners of cultural and media studies – Antonio Gramsci, Raymond Williams, Richard Hoggart and various members of the Birmingham School for Contemporary Cultural Studies. In different ways, each of these developed radical ideas about the 'work' that cultural texts perform in legitimising social relations, accentuating the need to examine texts in relation to ideological beliefs and structures. For example, Antonio Gramsci's neo-Marxist perspective alluded to the ideological role of the media in reproducing particular ideas about society. Hegemony, in his account, represented a struggle for moral, political and intellectual leadership, played out in the mass media as much as in the workplace or marketplace (M.J. Smith 2002).

The ability of texts to impose (and, on occasion, contest) 'common sense' understandings of the city thus provoked geographers to consider the ideological beliefs sedimented in both 'high' and 'popular' expressions of culture, including not just books and poetry, but also adverts, photographs, magazines and newspapers (Short 1991). Simultaneously, the adoption of terminology associated with *semiotics* (Box 2.2) became commonplace in geographical writing on text. Semiotics stresses that the meaning of language is, in effect, arbitrary, lacking constancy until it is given meaning through correspondence with other signs (Berger 1977). Written language, as a sign, signifies something only when it is 'placed' in relation to other signs. Moreover, it is argued that signification remains specific to a given audience, so that texts are potentially *polysemous* (i.e. may be understood in conflicting ways by different social groups). This implies that the meaning of text is not intrinsic, but is structured through social codes and conventions which encourage certain 'readings' of text (Slater 1998). In turn, these readings are seen to reproduce certain ideologies, values and philosophies as natural, universal and eternal.

Box 2.2 URBAN SEMIOTICS

Literally meaning the 'science of signs', semiotics can trace its origins to the work of both Charles S. Pierce and Ferdinand Saussure, who worked independently in the early years of the twentieth century developing a conceptual language and method for analysing signs. This method involves taking images apart and considering the signs as having two components: the *signifier*

continued

(a sound or image) and its *signified* (a concept). The latter is related to its real world *referent*: thus the word 'cold' (a signifier) can signify cold (as can the colour blue, or a picture of snow) and this refers to the feeling of actual experience of lacking warmth. What is apparent is that some signifiers work in different social and cultural contexts as they are *iconic* (for instance, a picture of snow); others are culturally specific and *indexical* as they rely on certain conventions of language and representation. In most studies, it is necessary to consider the relationship between signs, as the signifieds attached to certain signs get transferred to other signifiers, and often build to create elaborate *codes*: highly complex patterns of meaning that are common to particular society at a particular time (Aitken 2005). As such, semiotics provides an elaborate terminology for describing visual images and has been deemed a productive way of thinking about visual meanings. Its focus on ideology – and especially the codes which connect the meaning of signs to economic and political structures – means it is particularly useful for thinking through the social effects of representation (Rose 2001). Given this, significant attention has been devoted to the ideological construction and significance of urban spaces (Gottdiener and Lagopoulos 1986). This has involved semiotic approaches and techniques of reading the landscape to explore the social construction and power embedded in built environments. Much of this work has been influenced by Marxist traditions, as well as the writings of Baudrillard on signification and alienation. This is particularly evident in semiotic readings of spectacular spaces of consumption such as malls, festival marketplaces and leisure parks, which semiotic analyses reveal as illusory places of pleasure, leisure, hyper-reality and simulated 'elsewhereness' (Hopkins 1991; Shields 1991).

Key reference: Gottdiener and Lagopoulos (1986)

The idea that texts contain signifiers that send particular ideologically charged messages to different social and cultural groups was accordingly influential to those geographers seeking a more rigorous methodological and theoretical framework for the interpretation of text – Burgess and Gold's (1985) collection brings together a representative sample. Burgess's (1985) own work, for example, illustrates how newspaper reports served to perpetuate racism through the selective and unrepresentative reporting of inner city riots. In her account, the way that the text (and associated photography) encoded the story of the riots for its readers was indicative of the ability of a text to render an essentially biased and selective interpretation as 'common sense'. In the case of the rioting, this involved a consideration of the ability of the text to classify groups and individuals along racial lines, stereotyping non-white communities as inherently criminalised (see also K. Anderson 1991; Wall 1997). Here, the exclusions and narrative silences in the text are as important as what is included, with the selective and partial representation of people and place deemed a crucial means by which social inequality has been perpetuated (and justified); as cultural theorist Stuart Hall (1990: 156) argued, 'whoever controls information about society is, to a greater or lesser extent, able to exert power in that reality.'

While this focus on newspaper and media reporting may imply something of a division between literature-based explorations of sense of place and neo-Marxist accounts of media power, the two were in fact entwined within a reformulated cultural geography which focused on the cultural politics of everyday life. The arrival of this 'new cultural geography' paradigm was signalled by a flurry of publications (e.g. P. Jackson 1989; Anderson and Gale 1992; Barnes and Duncan 1992; Duncan and Ley 1993) which highlighted questions of geographical representation and began to spell out the importance of exploring the discourses immanent in a wide variety of written (and spoken) texts. Defining discourses as 'frameworks that embrace particular combinations of narratives, concepts, ideologies and signifying practices relevant to a particular realm of social action' (Barnes and Duncan 1992: 12), a key focus of the new cultural geography was the spatial meanings that are transmitted and reported through different domains and texts so as to reproduce power. The implication of such work was that while there may be varying representations of people or places circulating in different texts, collectively, the embedding of discourses in social life serves to constitute these people and places, reinforcing dominant social systems (Holloway and Hubbard 2001). In contrast to the forms of cultural geography associated with the Berkeley School, the new cultural geography was thus identified with cultural politics; namely, the ongoing struggle between different interest groups and cultures to promote their particular representations and knowledges (while obfuscating or destroying others).

The insistence that discourse does not simply reflect an already-existent social reality, but enters into the constitution of reality, is an important precept of post-structuralism (see also Chapter 1). Michel Foucault's (1980) ideas on power/knowledge were an important influence on post-structuralism: in his analysis, social categories of identification were not seen to be created by individuals (or structures), but by discourses. By way of an example, we might consider his work on medical discourse. Emanating from particular institutional sites (e.g. the hospital), medical discourse divides populations into sick and healthy on the basis of particular signs and symptoms. Further, it makes distinctions between various ailments and conditions, dividing the sick into those who require acute or secondary care. In his acclaimed work on *Madness and Civilisation*, Foucault (1967) likewise stressed that definitions of mental illness are temporally specific, with discourses of psychiatry having effectively brought certain categories of people into being, suggesting that their exclusion from the mainstream is necessary for their effective treatment and rehabilitation.

Disrupting any straightforward separation of text and a 'pre-discursive' real world, Foucault's ideas of developing a critical genealogy of knowledge became widely (though not uncritically) adopted by geographers seeking to expose the importance of text in shaping the contours of everyday life. In short, Foucault's ideas offered a fundamental challenge for those geographers who imagined that places (and people) have a real or essential existence outside the realms of language, encouraging a focus on the relations of discourse, knowledge and space (Barnes and Gregory 1997). Here, it becomes important to distinguish between methods of *content analysis* (which considers the surface or manifest content of text) and *discourse analysis* (a way of analysing texts that thinks about content in relation to its effects). Focusing on the representation of a city in a novel may well be interesting, but what has become more interesting for geographers is thinking about the role that that representation plays in creating urban identities. From this perspective, questions of what is true and false in a text are irrelevant; instead the focus is on the ability of the text to create reality through the 'invention' and documentation of difference.

Within a cultural geography that emphasises theoretical diversity, fluidity and flux, this focus on discourse has seen geographers grappling with post-structuralist ideas about the construction of society, considering the active work that text and image perform in constituting the people and places that they apparently only describe (Jones and Natter 1997; Rose 1997). Crucial here is Derrida's notion of *différence*, a term capturing the sense of difference inevitably written into any text:

> The sign represents the present in its absence. It takes the place of the present. We cannot grasp or show the thing, state the present, the being-present, when the present cannot be presented, we signify, we go through the detour of the sign. We take or give signs. We signal. The sign, in this sense, is a deferred presence . . . According to this classical semiology, the substitution of the sign for the thing itself is both secondary and provisional: secondary due to an original and lost presence from which the sign thus derives; provisional as concerns this final and missing presence.
>
> (Derrida 1991: 60)

Elucidating the relations of difference internal to language and other cultural codes, Derrida has expanded upon these ideas at length, suggesting things take their identity only from that which they are not. An example here is the notion of the city itself, which can be understood only in relation to what it is not (i.e. the countryside). Conversely, to depict a place as rural is, at one and the same time, to stress that it is unequivocally not urban. Meaning is thus relational, and language built on a slippery system of signification where the meaning of words is always deferred and never fixed.

Admittedly, post-structural understandings of text have not entirely replaced materialist and semiotic interpretations, but they have undeniably ushered in a new language of 'deconstruction' and 'destabilisation' (Laurier 1998). Simultaneously, they have encouraged geographers to extend the textual metaphor to encompass a wide range of cultural representations: music (see Leyshon et al. 1998), photography (Kinsman 1995), public art (T. Hall 1997) websites and the Internet (Crang 2002). Additionally, geographers have begun to explore some of the ways the discipline itself represents the world to others through varied forms of geographical exhibition and visualisation. One particularly prominent tradition here has been to consider maps not as 'ground truth' but rather as representations which impose an artificial order on the complexity of the world. The work of Brian Harley on the deconstruction of maps is influential in this regard. Adapting a broadly Foucauldian method, Harley thus explored the 'agency' of maps, demonstrating that 'the map works in society as a form of power-knowledge' (Harley 1992: 243). Seeking 'to show how cartography also belongs to the terrain of the social world in which it is produced', Harley (1992: 232) drew upon Foucault's notions of knowledge/power to suggest 'we have to read between the lines of technical procedures or of the map's topographical content' to expose the 'rules of cartography' – rules that are seen to be influenced by the rules 'governing the cultural production of the map'. These rules, he argued, are 'related to . . . those of ethnicity, politics, religion or social class, and . . . are also embedded in the map-producing society at large' (Harley 1992: 236).

Through his selective histories of mapping, Harley made the point that maps convey particular knowledges to meet the needs and requirements of the powerful. Having established this, another of Harley's tactics was to draw on Jacques Derrida's notion of deconstruction to search for the slippages and contradictions within maps that actually undermine their authority and reason (see Lilley 2002). In this way, Harley sought to detail the 'duplicity' of maps – a duplicity that many geographers and cartographers ignored by imagining their maps to be increasingly scientific and truthful representations of reality. This duplicity can be illustrated with reference to arguably one of the most famous maps in the world: the map of the London Underground (Plate 2.2). Basically unchanged since originally designed by Harry Beck in 1931, the London Underground map provides an easily comprehended schematic guide to the underground train lines that extend over twenty miles from the heart of the city to its suburbs. Colour-coded to enable instant apprehension, the map nonetheless provides a highly distorted and simplified map of underground topography, with Beck's schema eschewing any direct correlation with the layout of the city it represented: the Thames is straightened out; the extent of central London is exaggerated and the distances between different stations bear little resemblance to the actuality. As such, Pike (2002) argues that Beck's map is best understood as a *conceptual* space which provided an idealised vision of a London where social and spatial segregations were overcome by the modern technologies of the underground (interestingly, Beck was an out-of-work engineering draughtsman at the time, and conceived of the

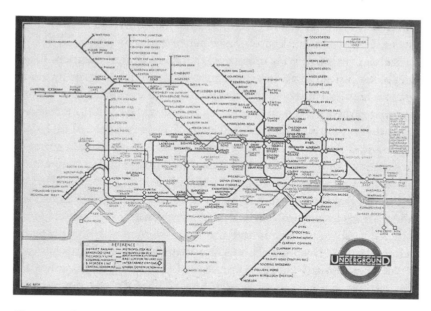

Plate 2.2 The London Underground map (courtesy of London Transport Museum)

city along the lines of an electric circuit board). Yet, by the same token, this conceptualisation has generated other effects, becoming an enduring icon of the city and shaping generations of Londoners' understanding of their place within the city.

Urban mappings such as that of the London Underground thus fulfil a deeper purpose than simply helping people orient themselves in physical space: they encourage us to conceive of the city in particular ways. Indeed, given that such maps emphasise certain places, people and flows, but suppress others, they encapsulate a particular 'way of seeing'. In the work of Henri Lefebvre (1991) on the production of urban space in capitalist society, it is suggested that these spatial *abstractions* typically perpetuate a rational, modern and technocentric viewpoint: what he termed the 'planner's eye view'. While alternative mappings are of course possible, Lefebvre suggested that, in the twentieth century at least, such representations of space conceived the city as a coherent, homogeneous whole which could be planned and organised so as to encourage capitalist development. This is mirrored in the tendency for maps and plans of the city to render invisible those things that planners, architects and politicians regard as 'out of place' in the city as conceived of from their 'rational', Cartesian and typically male perspective.

Rather than interpreting urban maps as objective representations based on rational, scientific procedures of surveying and mapping, many urban geographers have thus sought to explore how these spatial abstractions serve to perpetuate certain ideologies and ideals. Pike (2002) suggests Haussmann's fastidiously surveyed map of Paris (completed in 1853) represented the first example of such modernist abstraction given it provided the knowledge/power required to initiate a radical programme of urban improvement or 'surgery'. Similarly, Nead (2000: 22) alerts us to the way that Ordnance Survey maps of London were firmly embedded in the efforts to modernise London in the mid-nineteenth century, with 'the rules of measurement and the rules of society being mutually reinforcing'. As Nead details, modern maps effectively compartment-alised, classified and explained the logic of the metropolis, its aesthetics bequeathing a particular understanding of how the city ought to be planned and modernised. Poovey thus alleges mapping is a form of abstraction explicitly linked to a spatial geometry – 'a conceptual grid that enables every phenomenon to be compared, differentiated, and measured by the same yardstick' (Poovey 1995: 9). The same might be said of the plans for many of the new towns and state capitals of the mid-twentieth century, made at a time when confidence in modern planning precepts was at an all-time high. As James Scott (1998) relates, spatial order was manifest in remarkably aesthetic terms in these modern plans: simply put, they had to look regimented and ordered. Le Corbusier's utopian blueprint for the future city is therefore often cited as the quintessence of high-modernist urbanism, its influence on the design of new state capitals such as Brasilia, Ciudad Guyana and Chandigarh (as well as British New Towns, Swedish satellite communities and so on) demonstrating the affinity that emerged between state, capital and modernity in the mid-twentieth century (Holston 1991). In each, 'the order and certainty that had once seemed the function of a God was replaced by a similar faith in a progress vouchsafed by scientists, engineers and planners' (J.C. Scott 1998: 342). The plans which provided the template for urban development were thus exercises in modern abstraction rather than attempts to acknowledge the complex realities of urban life.

Of course, not all maps purport to represent reality. Del Casino and Hanna (2000) explore this when they write of the tourist maps of the city which are designed explicitly to steer visitors away from certain sites and towards others, dividing the city into visible and invisible attractions (see also Gilbert and Henderson 2002). As they contend, tourist maps may help to reproduce certain urban spaces as unique, exotic, exciting, leisurely or otherwise in contrast to the everyday spaces of work and home. By way of example, they focus on a German tourist guide to the nightlife of Bangkok (Thailand), which, through a series of suggestive icons and symbolic conventions, identifies certain clubs, bars and massage parlours as sites of sexual encounter. This type of mapping serves to steer tourists towards these sites, and away from others, producing a distinctively eroti-cised understanding of Bangkok's complex urban landscape. Yet Del Casino and Hanna (2000) go on to argue that such tourism maps do not only play a role in the production of tourism spaces, but also contribute to the reproduction of identities:

> Identities are defined and contested, and at times naturalized, through representational practices and individual performances. Many tourism maps include images of the people or 'hosts' who make tourism spaces unique, exotic and exciting. Tourists use these images to help them understand who these people are and how they are different from themselves. Some maps portray tourists engaged in the activities for which tourism spaces are produced, thereby confirming tourists' identities and guiding their reproduction of these spaces. In these and other ways, tourism maps contribute to the production of identity and can help us understand the relationships between identity, representation and space.
>
> (Del Casino and Hanna 2000: 25)

In a comparable manner, the guidebooks, postcards and brochures which are the stock-in-trade of the tourist industry can be read as texts which are implicated in the making of particular subjectivities and urban identities. Nevertheless, the meanings of such texts remain 'slippery', and there is much potential for dominant identities to be undermined or disturbed by alternative readings (as Del Casino and Hanna (2000) stress).

WRITING THE CITY

Clearly the notion of the text has been extended by geographers to encompass a wide range of images and representations. Indeed, the elasticity of the textual metaphor is illustrated by the fact that it has been extended to include the urban landscape itself. Following the lead of Denis Cosgrove, many cultural geographers now understand the landscape not so much as a portion of space, but rather as a particular way of representing the (visible) world through cultural forms (e.g. landscape art, perspectival architectural drawings, photographs and visual ensembles) that serve to make the environment legible, coherent and pleasing to the viewer:

> Landscape is in fact a way of seeing, a way of composing and harmonising the external world into a scene, a visual unity. The word landscape emerged in the Renaissance to denote a new relationship between humans and their environment. At the same time cartography, astronomy, architecture, surveying, land surveying, painting and many other arts and sciences were being revolutionised by application of formal mathematical and geometrical rules . . . Landscape is thus intimately linked with a new way of seeing the world as rationally-ordered, designed and harmonious.
>
> (Cosgrove 1989: 121)

The work of Cosgrove especially drew on the influence of Marxist cultural theorists and historians like Raymond Williams, John Berger and John Barrell to emphasise this way of seeing as inherently ideological. For instance, both Cosgrove (1984) and Cosgrove and Daniels (1988) explored eighteenth- and nineteenth-century traditions in landscape art, stressing that this mode of representation had close links with changes in land ownership and social relations in the countryside. In essence, such paintings were seen to incorporate an *iconography*, a set of symbols producing certain meanings of place in accordance with the interests of particular class groups. In turn, these symbols expressed a particular relationship between these groups and the landscape, so that the landscape was seen as the *natural* outcome of a particular mode of land ownership and husbandry. In short, the selective and stylised representation of landscape in art was interpreted as a statement of power, encouraging the preservation and maintenance of certain spaces and social relations (see Bender 1992; Daniels 1993; Nash 1996).

This type of interpretation, which brings social and cultural theory to landscape interpretation, offers a distinctive approach to understanding landscape, one which can trace its lineaments back to the cultural geography pioneered by Carl Sauer and the Berkeley School in the 1920s and 1930s (see P. Jackson 1989). In this tradition, the appearance of the natural landscape was seen to be transformed into a cultural landscape through the human practices and traditions indigenous to a particular area (providing 'a means of classifying areas according to the character of the human groups who occupy them' – Wagner and Mikesell 1962: 2). Less overtly radical in orientation than the iconographic tradition pioneered by Cosgrove and his colleagues, the Berkeley School's emphasis on mapping the distribution of material artefacts in the landscape continues to be an important influence, especially in North American geography. In the so-called 'new' cultural geography, however, the reading of landscape as caught up in political, social and economic power relations has encouraged more critical readings of landscape. The difference between these different traditions in cultural geography is subtle, and hotly debated (see W.J. Mitchell 1995), but in sum, it appears that the thrust of the new landscape studies is to consider landscape as part of a process of cultural politics, rather than as the outcome of that process.

This focus on the cultural politics of landscape has also provoked stimulating attempts to bring textual theory into dialogue with feminist and postcolonial theory. For example, postcolonial critiques of Western modes of representing and imagining the developing world have pointed out the importance of landscape in imperialist and colonial traditions (Kenney 1995; Neumann 1995; Phillips 1997). This notion of landscape as a highly classed, gendered and racialised way of seeing is also highlighted in many analyses of landscapes of national identity. Herein, the mapping of ideas of (for example) Englishness onto certain landscapes has been seen as central in processes of exclusion which serve to render ethnic, sexual and disabled minorities invisible in dominant constructions of national belonging (Kinsman 1995). In Daniels' (1993: 5) view, 'landscapes, whether focusing on single monuments or framing stretches of scenery, provide visible shape to the nation'. The connection between nationhood and identity as mediated through landscape has accordingly provided a key tradition in landscape studies, manifest in some lively and historically nuanced accounts of the role of visual culture in the legitimising of the nation state (Lowenthal 1991; Nash 1993; Matless 1998). While many studies emphasise the importance of rural and Arcadian myths in the construction of national identities, there have been several which have explored the role of particular built landscapes in consolidating national values and ideologies. Notably, Jacobs (1994) has considered the importance of Bank Junction (in the City of London) as a highly charged symbolic landscape, often perceived to be at the heart of the City and the heart of Empire (see Plate 2.3). When threatened with redevelopment in the 1980s, this site became the centre of a major public debate, in which luminaries such as the Prince of Wales invoked the national importance of the site, opposing this to the 'foreign' plans for the site's redevelopment. On a different scale, the layout of entire towns and cities may be read as expressions of national power, such as Canberra, designated as the new capital of Australia in 1908. Here, the state utilised the landscape to express and legitimise its power through the 'persuasive symbolism of landscape meanings' (Winchester et al. 2003: 71).

Likewise, it is clear urban public art generates ideological effects in so far as it mobilises meaning to sustain relations of political domination (see Miles 1998). The power of public art has certainly not been lost on those in positions of authority, with much research highlighting the role

Plate 2.3 Heart of Empire (Neils M. Lund, 1904) (permission of London Guildhall Art Gallery)

of public art in legitimating specific socio-spatial relations. For example, Atkinson and Cosgrove (1998) have asserted that the Vittorio Emanuele II monument in Rome was intended to embody particular notions of national identity, providing a 'memory theatre' through which the rhetoric of a united imperial Italy was to be conveyed (see Atkinson and Cosgrove 1998). Built in Beaux Arts style, this elaborate monument to the monarchy included multiple references to the history of imperial Rome in the recurring motif of the human body. Like many other monumental and sculptural works in the nineteenth century, this monument relied on the conventions of realist art and the adoption of a masculine 'way of seeing'. Moreover, its placement on a key ceremonial axis insured that this particular celebration of Italian identity was seen by many, and incorporated into national rituals of belonging and identification.

A ubiquitous feature of prominent public squares, official representations of heads-of-state and war heroes thus represent an important means of attempting to fix national memory – deciding not only who is remembered, but also the form these memories take. Related analyses suggest that monuments and built forms can imbue corporations, institutions and even individuals with power and prestige. Knox (1991) thus proposes the urban landscape represents a remarkable synthesis of 'charisma and context' (see also McNeill 2005, on 'skyscraper geographies'). As such, the townscape can be interpreted as 'written' by architects, developers and planners who operate within specific socio-economic context, its architectural styles and forms highly suggestive of the relations of culture, power and landscape. Domosh (1989) provides a useful guide as to how such a textual interpretation of the urban landscape might proceed, focusing on the sixteen-storey New York World Building, constructed in 1890 in Manhattan to house the offices of Joseph Pulitzer's The World newspaper. In this examination, she moves beyond a functional analysis of the building to explore the social and economic context in which it was constructed. Here, she suggests the building had to be tall not only to satisfy a demand for office space in a costly property market, but also to satisfy Pulitzer's personal ego; the French Renaissance Revival style demanded a level of respect, being fashionable among bourgeois classes at the time. Tacking between the shifting commercial imperatives of construction that informed the design and Pulitzer's flair for self-promotion, her reading thus alerts us to the multilayered task of interpretation:

> To understand the World Building necessarily involves one in several layers and types of interpretation – both functional (land values) and symbolic (need for legitimacy), of immediate relation to the object (Pulitzer's concerns) and very distant from the object (development of a class-based society). And all these different layers and types of explanation are related, although how these relationships work may not be so readily apparent.
>
> (Domosh 1989: 351)

While the importance of personal and corporate ego was crucial in the design and iconography of office buildings in Manhattan at the start of the twentieth century, King (2004) contends that such readings of architectural symbolism often need to be informed by a postcolonial perspective. This is the case both in the former colonies (as King shows in his writings on Old and New Delhi) as well as the 'imperial cities' – Vienna, Paris, London, Berlin – which were the command and control hubs of these empires (Driver and Gilbert 1999). Interrogating architecture and social spaces to reveal the colonial constructions and identities which they embodied has thus become an important task for those urban geographers determined to expose the ways in which colonial powers exerted their will over the colonised through the manipulation of space.

Such textured readings of the symbolism of the urban landscape indicate how the text metaphor has been stretched to encompass many urban phenomena. Here, it is apparent that geographers have also started to think about the ability of the contemporary urban landscape to promote particular messages through its appearance and design. Consideration of the way specific buildings promote particular cultural values and social rituals has accordingly been emphasised in studies such as David Ley's (1991) examination of the iconography of cooperative housing in Vancouver, Jon Goss's (1993) analysis of the signs and symbols of the West Edmonton shopping mall and Sharon Zukin's (1995) work on Manhattan, while a more general consideration of the staged inauthenticity of the postmodern urban landscape has been offered by David Harvey:

> We can identify an albeit subterranean but nonetheless vital connection between post-Fordism and the postmodern penchant for the design of urban fragments rather than comprehensive planning, for ephemerality and eclecticism of fashion and style rather than the

search for enduring values, for quotation and fiction rather than invention and function and, finally, for medium over message and image over substance.

(Harvey 1989b: 13)

For Harvey, the depthlessness of the contemporary city, with its emphasis on surfaces and signs, is evidentially a means by which capitalism has sought to both attract and distract, transforming spaces of production into spaces of postmodern play and capricious consumption. Disney-esque architectural assemblages are thus a familiar component of city redevelopment schemes, their playful forms obscuring their 'real' role as spaces of capital accumulation and real estate investment.

Ostensibly combining Marxist political economy with post-structural ideas about the instability of meaning, Harvey has offered some stimulating and influential interpretations of the new urban forms emerging in the post-Fordist city (see Chapter 1). Yet, for some, Harvey's continuing emphasis on the imperatives of (disorganised) capitalism lessens the value of his work. Nonetheless, his attention to the commodified signs, styles, metaphors and images is richly suggestive of the importance of signs and symbols in the consumer city. Here, there are significant links to several accounts of postmodernism – particularly that of Jean Baudrillard – which stress that the consumption of signs has become everything, their relation to real or authentic referents no longer an issue. Slipping into the seductive grasp of signs and simulacra, a widely noted tendency has been for consumers to buy into spaces offering multiple opportunities to get lost in this melange of signs and images. In Baudrillard's work, these are described as hyper-real because of the way they feign authenticity; examples include those contemporary shopping malls, theme parks and heritage centres that offer commodified fantasy versions of other places and times (see Shurmer-Smith and Hannam 1994). An example here is the hyper-real Irish bar, a familiar sight in cities around the world. Replete with draft Guinness, Gaelic football memorabilia, former Irish bank notes and post-cards, these bars offer a version of Irishness which is, conversely, more 'Irish' than that found in pubs in Ireland itself. Like the spaces of cinema, computer games and TV drama, these hyper-real pubs (and shopping centres, heritage parks and theme parks) are representations, at once real and imagined. In the final analysis, by demolishing any easy distinction between image and reality, work on the aesthetics of the postmodern city has therefore underlined the importance of reading space like a text, its symbolism being created *intertextually* via the meanings of other texts.

REWRITING THE CITY

While some geographers still insist that particular cities have an essence that can be identified and distilled, the majority would accept that cities are in fact heterogeneous and variegated entities whose spatiality escapes any attempt at mimetic summation and representation. Notwithstanding this, urban politicians and policy-makers constantly seek to do just that, offering up a boosterist narrative that seeks to define the character – perhaps even the *spirit* – of a city. In the post-industrial West, this narrative is promoted through a plethora of leaflets, websites, CD-ROMs, promotional T-shirts, postcards and videos, each of which extol the virtues of a given city as a great place to live and work. Arguably, such civic boosterism is most entrenched in the United States, where urban quality of life and city league tables are widely and hotly debated (see McCann 2004a), but it is in those cities where deindustrialisation wrought widespread disinvestment and decay in the 1970s and 1980s that the orchestration of place promotion has been most vigorous. In particular, it is the rust belt of the United States where cities are seen to have pioneered *city branding* (Box 2.3). In such cities, promotional strategies designed to enhance city image have been deemed essential in developing new markets, attracting new investors and preventing the exodus of affluent urban citizens.

Box 2.3 CITY BRANDING

In an era where city governors have become more entrepreneurial and business-minded in their outlook, it is perhaps not surprising that the concepts and language of marketing have infiltrated the realm of urban politics. One element of this is that city governors and promoters often speak of a distinctive urban *brand*. A brand can be defined as 'a mixture of tangible and intangible attributes, symbolized in a trademark, which, if properly managed, creates influence and generates value' (Clifton and Maughan 2000: xvi).

Marketers suggest that the brand is more important than the product that is being sold, and believe that communicating the core values of the brand is the key to generating customer loyalty and brand recognition. Normally developed after an intensive scrutiny of a city's positive and negative attributes for inward investors and tourists, urban branding is designed to bequeath the city with a recognisable and consistent identity which people will immediately associate with a particular city. Kavaratzis and Ashworth (2005) suggest that such branding is often one-dimensional, and seeks to promote one particular facet of urban identity so as to develop a niche in global flows of investment and tourism. This may be through the forging of a connection between a particular city and a personality (Joyce's Dublin or Gaudí's Barcelona), stressing the key contribution of a major landscape or Prestige Project (such as the Guggenheim museum in Bilbao) or highlighting a major cultural or sporting event (the Venice Biennale, the Monaco Grand Prix or the Edinburgh Festival). Naturally, this is tied into an effort to attract particular types of consumption or investment (e.g. Birmingham's branding as 'Europe's meeting place' was a self-conscious attempt to promote the city to high-spending business tourists and conference goers). While possessing a strong and distinctive city brand may help a city position itself on a global stage, there are also inherent dangers in promoting one facet of the city's identity over others: all brands have a sell-by date. The constant monitoring, revision and reworking of brand identities are therefore recognised to be crucial if cities are to maintain their competitive advantages over rival brands.

Key reference: Gold and Ward (1994)

What is clearly evident here is that contemporary forms of place promotion are not simply attempts to advertise the city. Rather, the intention is to reinvent or rewrite the city, weaving myths which are designed to position the city within global flows of urban images and representational practices. Short (1996) suggests the key myths here are that the city is multicultural (and accepting of difference); that it is environmentally friendly, rich in cultural attractions and supportive of new investors – or, to put it another way, that it is simultaneously a fun city, a green city, a pluralist city and a business city (see also Short and Kim 1998).

Flowerdew (2004) presents an interesting case study of how city authorities attempt to invoke such myths. Focusing on Hong Kong he details how senior public and private sector representatives came together to consider how to develop the former British colony as a city 'enjoying a status similar to that of New York in America and London in Europe'. A report published in 2000 – Bringing the Vision to Life: Hong Kong's Long-Term Development Needs and Goals (Hong Kong Special Administrative Region 2000) – underlined the goal of developing Hong Kong as 'Asia's World City'. The five core values at the heart of the Brand Hong Kong programme were stated as follows: progressive, free, stable, opportunity and high quality. These themes were discursively constructed not only through the reports and consultation documents that set out the vision for Hong Kong, but also the many promotional videos and brochures seeking to represent Hong Kong as a site for inward investment, tourism and trade. Flowerdew describes the content of one of these (the Gateways and Portals video 2001) thus:

> We have images representing business activity, with an emphasis on the use of technology such as computers and mobile phones; media production technology; trade (the container port); vibrant night life (neon signs); a modern transport system (the underground rail system); the high speed railway that links the centre of Hong Kong to the new airport; the airport itself; the rule of law (lawyers in wigs and gowns, the Legislative Council and its statue of Justice); a free press (a collage of the logos of local and international print media); education (students celebrating their graduation); global links (the Chief Executive meeting Mickey Mouse [Hong Kong has an agreement for a Disneyland to be built there]); and modern architecture (scenes of skyscrapers and other modern buildings). The video culminates with the gradual appearance of the visual symbol of the Brand Hong Kong, a stylized dragon on the left side of which can be deciphered the words HONG KONG in English and Chinese. This final image coincides with the final line 'Hong Kong: Asia's World City'.
>
> (Flowerdew 2004: 595)

To some extent, Hong Kong's attempt to identify itself as a prime site for global investment and trade has been hampered by other images and associations (such as the Asian financial crisis of 1997, and the outbreak of SARS in 2001) which are not dwelt on in such promotional texts. Similarly, Short et al. (1993) note that many cities remain encumbered with negative images of industrial decline and dereliction which are not easily reversed: it is difficult for some of the heavy engineering cities of the former East Germany to promote themselves as unspoilt post-industrial cities given that they are routinely depicted as having been heavily polluted.

Reversing inherited images of urban decay, inequality or pollution is thus essential in a global economy where 'image is everything'. Yet even when few of these inherited images exist, urban marketeers may still face an uphill struggle in selling their city as ripe for investment. For instance, some cities are simply too small or anonymous to be widely known among those who make international invest decisions. One such example of a city with a 'weak' image is Leicester in the English East Midlands. Although one of England's most ethnically diverse cities, research com-missioned by Leicester City Council suggested that people know little about it compared with cities of similar size (e.g. Glasgow, Liverpool, Nottingham or Bristol). In part, this perception was shared by Leicester residents, who displayed little local pride and typically described the city as distinctly ordinary (Landry et al. 2004). On this basis, the Council launched 'Leicester Revealed' – a seven-year initiative to improve the image and perceptions of Leicester. This controversially kicked off in 2003 with street posters bearing slogans like 'Boring, Boring Leicester' and 'Leicester . . . Nothing to Shout About!' (see Plate 2.4). Seeking to acknowledge negative images and addressing them head-on, this campaign provides an interesting example of how place marketing can provoke local people to take more pride in their city (in this instance, not entirely successfully).

Nevertheless, such promotional campaigns can backfire if local populations do not identify with the messages being promoted. Much geographical scrutiny of place marketing has thus focused on the way 'official' representations occlude certain place identities, alienating specific communities. For instance, Boyle and Hughes (1995) contend that Glasgow's reinvention as European City of Culture (1990), pioneered through the 'Glasgow's Miles Better' campaign and the promotion of local architects and artists like Charles Rennie Mackintosh, was one that

Plate 2.4 Poster used in promoting Leicester (courtesy of Leicestershire Promotions)

alienated much of the city's population. Specifically, it was seen by many on the left as a denial of the city's working-class heritage because of its failure to celebrate (for example) the contribution of the ship-building industry to the local economy, excluding industrial representations of the city in favour of a glossy, post-industrial vision. Subsequent counter-representations by leftist and labour-based groups sought to redress this balance, aiming to shatter the newly manufactured image of the city in favour of one that better represented the lifestyles and experiences of Glasgow's indigenous populace.

Equally, in other cities it is possible to discern that the hyperbole of place marketeers often ignores the contribution of specific social groups and neighbourhoods to the city in favour of a 'soft-focus' imagery designed to appeal to middle-class consumers. What this demonstrates is that representations are rarely innocent, being enmeshed in a complex cultural politics by promoting certain senses of place in favour of others. Brian Graham (2002: 101) therefore makes a distinction between two parallel cities: the 'external city', which is known through dominant representations of tourist icons and spectacular landscapes, and the 'internal city . . . concerned with social inclusion and exclusion, lifestyle, diversity and multiculturalism'. The tension between these is thus crucial in understanding the process of city promotion, with careful orchestration needed to create meaningful and believable place myths around which local interests can unite.

Nevertheless, as many urban geographers have pointed out, place promotion is rarely limited to the launch of a new advertising campaign, and frequently goes hand in hand with the creation of a new urban landscapes. Indeed, the construction and promotion of iconic new urban landscapes, typically in derelict or waterfront areas, has been an almost universal response to deindustrialisation in British and American cities, following the success of the rejuvenation of Baltimore's waterfront in the 1960s. Such redevelopments frequently centre on a 'flagship' project, a term commonly applied to a pioneering, large-scale urban renewal project such as a cultural centre, conference suite or heritage park designed to play an influential and catalytic role in urban regeneration. Examples include the redevelopment of London's Docklands, New York's Battery Park, Barcelona's Olympic Marina, Birmingham's Brindley Place, Paris's La Defense, Lisbon's Expo Centre, Berlin's remodelled Potsdamer Platz, and so on. Inevitably, this redevelopment and 'repackaging' of urban districts into areas of (and for) consumption is heavily promoted by urban governors, in effect becoming a representation of the city in its own right (Hubbard 1996). The appearance (and *scale*) of these new packaged landscapes has been proposed as a crucial instrument in the production of a city's images: McNeill (2005) identifies the ongoing race to build the world's tallest building as evidence of the importance of architecture in city promotion.

Of course, there is no guarantee that such schemes, or the publicity they generate, will create additional revenue for the city, and some are financial disasters. This means most have to be underwritten by the state and financed by private sector capital. Frequently, this means the public sector bearing the losses while private sector investors (including large corporate developers and financial institutions such as investment banks and insurance fund managers) reap the benefits. Private capital is thus attracted by the reduced risk implicit in such public–private partnership initiatives, creating a growing 'demand' for similar developments (Gaffikin and Warf 1993; Swyngedouw 1997). By the same token, it is notable that most schemes are targeted to a new global elite, with these urban development schemes supplying the working, living and leisure spaces for those who actually have little commitment to a particular city and live in the essentially self-contained 'islands of the cosmopolitan archipelago' (Bauman 2000: 57; see also Sklair 2005).

In many cases these schemes involve the transformation or gentrification of redundant or 'low grade' sites with redevelopment potential in or around the older quarters of city centres. This has led several commentators to implicate place marketing in a process of corporate-led gentrification, with the re-aestheticisation of the city centre accompanied by the development of 'playscapes' catering to the affluent (Hackworth and Smith 2001; MacLeod and Ward 2002). Hence, while such strategies of city promotion unquestionably improve the aesthetic appearance of downtown districts, when viewed through the lens of political economy, this rewriting of urban space is interpreted not as an end in itself, but the means to an end, namely, the accumulation of capital (Harvey 1989b; Swyngedouw 1997). Consequently, many urban commentators have found it difficult to be sanguine about the impacts of place promotion for those marginalised by capitalist processes. For example, Leitner and Sheppard (1998) have assembled damning empirical evidence to suggest boosterist policies merely deepened existing socio-economic polarities.

As they detail, while growth-promoting strategies created images of prosperity in such declining cities as Pittsburgh, Cleveland and Glasgow, they did not redress such problems as a shrinking number of employment opportunities, neighbourhood decay, and fiscal squeeze. Indeed in some cases they exacerbated them. Leitner and Sheppard (1998) hold up Pittsburgh as emblematic: heralded in the media as a success story because of its revitalised downtown, they point out that it nevertheless had the second-highest black poverty rate among the twenty largest metropolitan areas (36.2 per cent) in 1990, and the sixth highest male unemployment rate (12.1 per cent). For them, this underlines the more general argument that urban pro-growth agendas inevitably intensify social and territorial inequalities within cities.

Despite such critiques of place promotion, almost every Western city (and many non-Western) continues to pump out an endless stream of marketing material and hyperbole. What is particularly notable about this is the remarkably uniform narrative employed by city governors, with each city represented as 'conveniently located', boasting a wide range of cultural amenities and populated by a willing and well-educated workforce. Ultimately, this serial replication of urban representation appears to bequeath weak competition and an overcrowded marketplace (McCann 2004a). Hence, while city promotion is commonly understood as a means to achieve competitive advantage in the global economy, as well as enhancing civic pride and social inclusion, the contradictions and ambiguities of this process are clearly etched in the boosterist rhetoric of place promotion (see also Chapter 5).

CONCLUSION

Despite occasional attempts to examine the impact of city images on spatial behaviour, geographers have traditionally eschewed consideration of urban representation, preferring to describe and explain the formation of 'real' urban landscapes. As a reaction to this, the 'representational turn' has had important effects, bringing questions of culture, power and representation squarely within the ambit of urban geography. As we have seen, metaphors of reading and writing have ushered in new ways of thinking about urban texts: the city is *deconstructed* to reveal its ideological forms; its *iconography* is teased out through careful contextual analysis; urban mappings are unpacked to reveal specific geographical *imaginations* of the city. Clearly, the theoretical influences here are wide and varied, taking in socio-semiotic frameworks rooted in political economy as well as post-structural perspectives that emphasise that there is 'nothing beyond the text'.

Reflecting on this representational turn, it is important to stress its ambivalent effects. On the one hand, an emphasis on questions of culture and image has alerted urban geographers to the limitations of analyses which ignore the vital importance of language and text in mediating urban life. On the other, it has been criticised for distracting from 'real' problems of poverty, inequality, deprivation and crime in cities (Peet 1998; Hamnett 2003). While many of these criticisms are misplaced, and potentially easy to allay, the idea that geographers were being led into a position where all they could do would be to offer 'readings of readings' was one that triggered a number of reactions. Hence, while this trope of city-writing (the idea of city as text) is widespread in contemporary urban geography, it is far from hegemonic, being accompanied by other important conceptualisations of urban space.

FURTHER READING

Winchester et al. (2003) Landscapes: Ways of Imagining the World provides an excellent introduction to cultural geography, and includes many examples relating to the representation and mapping of the city. Richard Jackson's (1990) Cultural Geography remains the key text on language, space and representation, while Shurmer-Smith and Hannam's (1994) Worlds of Desire, Realms of Power devotes much attention to post-structural and feminist understandings of landscape. More explicitly urban in orientation is Paul Knox's (1993) The Restless Urban Landscape, which provides some absorbing textual readings of urban spectacle, while many chapters in Duncan and Ley (eds) (1993) Place/Culture/Representation focus on case studies of gentrification and residential landscapes. For those conducting their own studies of urban representation, an extremely useful guide to visual methods and semiotics is provided by Gillian Rose's (2001) Visual Methodologies.

3

THE EVERYDAY CITY

To think about the city is to hold and maintain its conflictual aspects: constraints and possibilities, peacefulness and violence, meetings and solitude, gatherings and separation, the trivial and poetic, brutal functionalism and surprising improvisation.

(Lefebvre 1996: 53)

While Chapter 2 focused on the broad gamut of work considering the city as text, it is evident the 'cultural turn' in the social sciences has inspired other takes on urban culture. One such take is that which sidesteps questions of text and textuality to focus on the *textures* of the city, not least those created through the social practices of the everyday. Central to this perspective is the idea that the majority of social action in the city is unmediated: that is to say, it does not appear to rely on representational knowledges or images of place. Everyday life in cities is, after all, something that cannot be adequately prepared for: no matter how carefully scripted, urban life has a tendency to surprise, and we are constantly forced to improvise and adapt to events as they unfold around us. This is something particularly evident in the more banal and routine aspects of urban life (for instance, the way we walk, talk, drive and generally negotiate our way through the city streets), yet it is clear that everyday life in cities is characterised by all manner of practical adaptation. What is evident here is that while individuals seldom have much control over what goes on in cities, they creatively improvise to open up 'pockets of interaction' in which they can assert and express themselves (no matter how fleetingly or inconsequentially) (Thrift 2003b: 103).

For some commentators, a focus on practice is deemed necessary to provide a corrective to the type of ideas considered in Chapter 2. For some, the textual metaphor has been stretched too far, neglecting issues of materiality in favour of a focus on a representation that is insubstantial and, perhaps, unreal:

> If all culture, and all the world, becomes a matter for representation, then we may lose purchase on the differences of material substance, whether that material is concrete, earth, paper, celluloid, and similarly, the power of the textual metaphor may be lost through over-extension.
>
> (Matless 2000: 335)

For such reasons, the emerging body of work focusing on the practical negotiation of the city can be interpreted as something of a reaction to the perceived limitations of approaches that prioritised examination of the imagined city rather than its 'material' counterpart. Equally, it can be read as an attempt to *revitalise* an urban studies where analysis of representational space had taken precedent over lived space. By the late 1990s, a well-established critique of 'representationalism' in general – and cultural geography's landscape fixation in particular – had emerged, centring on the criticism that it 'framed, fixed and rendered inert all that ought to be most lively' (Lorimer 2005: 85; see also P. Jackson 2000; Wylie 2002). Critics warned against 'futile exercises of deconstruction' and the 'verbal games of postmodern theorising' (Castells 2001: 404), suggesting questions of representation distract from the real business of urban studies: producing socially relevant knowledge that can help alleviate poverty, injustice and violence (Hamnett 2003). Cresswell (2003) voiced another significant concern, namely, the exclusivity of much landscape inquiry rendered it inaccessible: in his view, geographical study can be made simultaneously meaningful, popular and political only through a closer engagement with everyday practice.

As we shall see, many of the criticisms of geography's cultural and representational turn have been misplaced. Nonetheless, such criticisms echo those voiced elsewhere in the social sciences, where a concern that the 'cultural turn' has run its course has triggered a search for new theories which focus on practice and action (Miller 1998). Ironically, this 'practical turn' has involved a re-engagement with the work of Marxist humanists including Henri Lefebvre, Guy Debord, Michel de Certeau and Michel Bakhtin, as well as more recent theories of social action and 'performativity' (see Highmore 2002a). In this chapter, I therefore explore how geographers have grappled with the life of the everyday city, focusing on questions of embodiment, action and practice. Sketching the contours of a putative 'non-representational theory', I thus outline some of the distinctive ways that contemporary geographers are approaching questions of practice, and, on this basis, consider what this tells us about the nature of urban life.

CITIES FROM ABOVE AND BELOW

At first, it appeared simple: to gain an overview of Paris, one needs to find a suitable viewpoint. But where from? The top of the Montparnasse tower? No – the crowds there would distract. The top of Montmartre (where one would have the advantage of not seeing the Basilica Sacré Coeur)? Perhaps, but the view would be too oblique there. Maybe a satellite photograph? But then only one plan would be obtained . . . The balcony of the office of the Mayor of Paris, in the Hotel de Ville? An empty and cold place, surrounded by ugly fountains: one would see nothing of the vitality of the metropolis. So, is it impossible to comprehend the city? Not if we keep moving! Let us circulate, and then, suddenly, Paris will become gradually visible.

(Latour and Hermant 2001, my translation)

As Bruno Latour and photographer Emilie Hermant imply, the search for the essence of a city is one that impels urban researchers to look at the city from different vantage points and viewing angles. Traditionally, it has been thought that one of the best ways of comprehending the city is to look at it from afar and above: the mythical 'god's eye' view of cities which informs all manner of maps, plans, models and abstractions of urban space (Vidler 2002). As we saw in Chapter 2, the pursuit of totalising visions of the city has led to the powerful seeking ways of surveying and mapping cities from the air, from the earliest surveys by balloons to the use of photogrammetry and satellite imagery. Such aerial perspectives allow one to perceive and represent the city as a totality which can be easily and immediately comprehended (and, by implication, ordered).

The appeal of such an overview of the city was famously captured by de Certeau (2000: 101–102) when he wrote of the 'violent delight' of travelling to the top of the World Trade Center in New York, high above the roar of the 'frantic New York traffic'. The popularity of viewing platforms and observatories in some of the world's tallest buildings is testament to the thrill of seeing the city from this vantage point (see Plate 3.1). Adopting some of the metaphors of reading and writing we explored in Chapter 2, from this perspective it is possible to read the city as a vast and static panoramic text which stretches out before the viewer. While the pleasures of the tourists who throng around the telescopes and troposcopes on observation decks are normally quite innocent, it is from similar vantage points that the powerful have been able to project their mastery over the urban scene. The appeal of this 'panoptic' view for those who would seek to order and govern our cities is obvious: it allows them to conceive of a city in which many of the things which potentially disturb their conception of an ordered city can be conveniently ignored: from above, one cannot register the rubbish that piles up in the city's back alleys, hear the constant noise of traffic and people, or smell the fumes that choke the streets below. Crime, deprivation, death, loss, anger and pain all cease to have any existence in a city that is rendered as a visual ensemble rather than a living, breathing entity.

Hence, as de Certeau spells out, the view from above may exercise a strong appeal, yet it may lead us to (literally) overlook some of the most important facets of city life. For de Certeau (2000), 'the story begins on ground level with footsteps' – the activity of thousands of walking bodies constructing the city at street level through a multiplicity of footsteps which can never be added and which never comprise a series. Continuing with a textual metaphor, he argues that pedestrians 'write' the city without 'reading' it. By this, he implies that when we walk in the city, we are indifferent to representations, not least the panoptic knowledges of those who would control our cities. In effect, the movements of people (and traffic, and animals, and goods . . .) construct a different urban text which cannot be apprehended from afar and above. To ignore this text is thus to ignore a key dimension of urban life. As Benjamin (1985: 50)

Plate 3.1 The Seattle Sky Needle, opened for the 1962 World's Fair, taps into our desire to see the city 'from above' (source: Bernard Mayo Rivera, United States Coast and Geodetic Survey)

stressed: 'the airline passenger sees only how the road pushes through the landscape, how it unfolds according to the same laws as the terrain surrounding it: only he who walks the road on foot learns of the power it commands.'

There are thus important connections between scopic regimes and the exercise of power: as Rose (2003: 213) points out, visualities always have 'their foci, their zooms, their highlights, their blinkers and their blindnesses'. Putting it bluntly, Miles (2002) argues that official accounts of city life (maps, census returns, plans etc.) write the city from the point of view of an authoritative, privileged male. He argues that this 'view from above' unifies disparate elements of urban form, reducing 'human participants in its spectacle to a role equivalent to the figures in an architectural model.' Repressing the agency of those who live and experience the city at 'street level', the implicit danger here is that we get a distorted record of a city's redevelopment: a story told 'by intellectuals and for intellectuals' which ignores the complex ways that 'ordinary' citizens engage with, and change the city in the realms of everyday life. As Thrift (1993: 10) contends, the result of such writing strategies is to perpetuate 'a particular form of urban theory which sees the city as the stamp of great and unified forces which it is the task of the theorist to delineate and delimit.' In his estimation, this creates distinctly modern accounts of modern cities, and dispels the 'magical' elements of urban life, not least the extraordinary capacity for urban dwellers to change the city through everyday practice (see also Pinder 1996).

This emphasis on everydayness is nonetheless problematic (see Holloway and Hubbard 2001). Elaborating, Highmore (2002a: 16) proposes that the everyday is 'a problem, a contradiction and a paradox', being simultaneously 'ordinary and extraordinary' because it is both a collection of repetitive and banal actions which reproduce the status quo, yet is also a site of resistance, revolution and ceaseless transformation. Everyday life has consequently captured the attention of many cultural theorists and activists, who have sought to articulate its ambivalence. Among these thinkers, it is Walter Benjamin who probably did more than any other to elucidate the equivocality of the everyday city. Most famously, his unfinished Arcades project sought to understand how Paris became an industrial capitalist city in the nineteenth century through a series of notes and essays inspired by his wanderings through the shopping arcades of the 1930s. As was described in Chapter 2, Benjamin used the poetry of Charles Baudelaire to exemplify the sense of magic and loss that was experienced by those living in the city as it underwent rapid and spectacular urban redevelopment. As a wealthy 'man of the streets' (or flâneur – see Box 3.1), Baudelaire treated the streets of Paris as his own personal realm, visually consuming its sights and sounds and using them to inspire his art and poetry. His poetry captured both his desire for the new experiences that could be bought and sold in the cafés, arcades and theatres of Paris, as well as his sense of disgust that life had been reduced to an industrial process of production, advertising and consumption. For Benjamin, Baudelaire captured the essence of modernity – and signalled the key contradictions evident in everyday urban life. As such, Baudelaire became Benjamin's alter ego – a figure whose intoxication with Paris reflected Benjamin's own dictum that each moment provides an opportunity for profound illumination (see Merrifield 2002).

Benjamin thus delved into both the arcane and mundane elements of urban existence in his efforts to apprehend the actuality of everyday life. Antiques, books, trash, dioramas, food, films, prostitutes and rag-pickers are all discussed in the unfinished Arcades project. Here, and in his other works (especially One Way Street 1985, originally published in 1927), he favoured juxtaposition and montage to draw attention to the contradictions of everyday life. This allowed him to highlight both the 'cancerous' commodity fetishism that intruded into every facet of urban life while still expressing the sheer vitality and possibility of the city (Benjamin 1999: 872). Further, he believed this form of dialectic imagery would 'break the spell of the capitalist dream' and give rise to new forms of life (Highmore 2002b).

Through his attention to the poetics of the street, Benjamin demonstrated that Marx's discussion of capitalist commodity forms could be brought down to street level – and, conversely, that the everyday city could be thought about critically. Similar ideas about the revolutionary potential of street life are offered in the work of Henri Lefebvre. Renowned as a philosopher of the everyday, Lefebvre produced a remarkable series of publications that explored how the lived worlds of people (and their sensual and sexual desires) had been gradually colonised ('papered over') by the forces of capitalism (see Lefebvre 1972). In his final works, Lefebvre argued that capitalism had survived and flourished by producing and occupying space, suggesting that each society produces a space suited to its own perpetuation. In effect, this superseded Marx's historical materialism (where class conflict is theorised as the basis of social

Box 3.1 THE FLÂNEUR

While the flâneur is not the only male who used the modern city as a site for observation (e.g. see Farish 2005 on the detective), the flâneur has become a key figure in the histories of urban space and an important trope in city-writing. In essence, the flâneur exists both as a real figure who inhabited the modern city and a metaphor for a particular mode of visual apprehension. In relation to the former, the flâneur was a fashionable man of leisure for whom the streets of the city effectively served as a living room, place of work and source of artistic inspiration. Inevitably a member of the upper classes, the flâneur and his close relations, the rambler, dandy or macaroon, seemingly had no need to work, and instead spent their time promenading in the city. At one and the same time, they were a quintessential part of the urban scene yet also remained distanced and aloof. Their privileged role as men of the streets thus afforded them a distinct vantage point looking at the city and re-presenting its sights and sounds in their poetry, art or writing. As an ideal of urban exploration and representation, many have followed in the footsteps of the flâneur. However, the fact that the flâneur is a gendered figure whose gaze objectified women and allowed them little agency has often been commented on (see especially E. Wilson 1991). It is thus significant that there is no female equivalent of the flâneur, with the 'women on the streets' (i.e. street sex workers) having been subject to the predatory advances of the flâneur. Weighing up these arguments, some have suggested that romanticising the flâneur is dangerous, and that flaneurialism has passed its sell-by date as a mode of experiencing and writing the city.

Further reading: Jenks and Neves (2000); Rendell (2002)

change) with a geographical materialism that focused on spatial conflict. Lefebvre accordingly outlined the importance of urban space in effecting the transition from classical and feudal society (typified by 'historical space') to a capitalist society (characterised by abstract space) by routinising and legitimising the rhythms of everyday life:

> What space signifies is dos and don'ts – and this brings us back to power. Space lays down the law because it implies a certain order – and hence also a certain disorder. Space commands bodies. This is its *raison d'etre*.
>
> (Lefebvre 1991: 121)

Lefebvre dismissed the idea that urban space simply exists, emphasising its social production by distinguishing between three forms of space: spatial practices (the routines that constitute the everyday), representations of space (the knowledges, images and discourses that order space) and spaces of representation (which are created bodily). He suggested that the perceived world of spatial practice (life) was held in a dialectic relation with representations of space (concepts), but that this dialectic could be transcended via creative bodily acts ('life without concepts'). Outlining the interplay between these forms of space, Lefebvre thus developed a trialectic in which all three are held in tension, albeit he suggested that the excessive energies of the body act as a mediator – and potential transformer – of the relationship between these different dimensions of space (Merrifield 1995). Hence, while the (labouring) body is central to the working life of the city, Lefebvre suggested it has the capacity to create cities in which a wider range of desires – including sexual passions – could be realised. Significantly, much of his later writing focused on the revolutionary potential of everyday life by focusing on the love, poetry and games which punctuate the everyday life in moments of 'festival' (Simonsen 2005).

Though Lefebvre expressed considerable faith in the ability of people to create cities which would fulfil their bodily needs and desires, his fear was that the abstract representations of space deployed by architects and planners over-coded 'spaces of everyday practice' by 'de-corporealising' the city. Discussing the world of images and visualisation, he suggests that bodies are effectively 'emptied out', and stripped of their life and warmth by technocratic planning discourse that focuses on the functional order of urban space. An example here is the designation of shopping spaces as just that – spaces for consumption (and not eating, skating, dog-walking or hanging out – see Plate 3.2). This resonates with Michel de Certeau's (1984: 59) discussion of the strategies (of power) that constitute a mode of administration. In his words, these strategies are those practices of ordering that repress 'all the physical, mental and political pollutants'. De Certeau argued that this is mirrored in the tendency for architect-planners to render invisible those things they regard as 'out of place' in the city as conceived of from their 'rational', phallocentric, Cartesian perspective. This 'planner's eye view' thus exemplifies the discursively constructed 'representations of space' that Lefebvre felt were vital in ensuring the domination of capitalist space (based on exchange values) over fully lived, spontaneous and creative space (based on use values).

McCann (1999) suggests that Lefebvre's attention to everyday life makes his work particularly relevant to analysis of public spaces, including streets, parking lots, shopping malls and parks. Although these are places often rationalised and planned as spaces of consumption, studies suggest that their occupation frequently challenges this dominant representation. This can be illustrated with reference to rituals of skateboarding, an embodied practice which often challenges the dominant conception and occupation of space:

> Skaters' representational mode is not that of writing, drawing or theorising, but of performing – of speaking their meanings and critiques of the city through their urban actions. Here in the move-ment of the body across urban space, and in its direct interaction with the modern architecture of the city, lies the central critique of skateboarding – a rejection both of the values and of the spatio-temporal modes of living in the contemporary capitalist city.
>
> (Borden 2001: 25)

As Borden (2001) details, the skater's engagement with the city may be conceived as a 'run' across its terrains, including encounters with all manner of diverse objects and spaces: ledges, walls, hydrants, rails, steps, benches, planters, bins, kerbs, banks and so on. In this sense, skaters see the city as a set of objects or surfaces which offer a series of challenges or experiences. This tendency to view the city in this way contrasts markedly with the architect-planners' view, which conceives of the city not as a

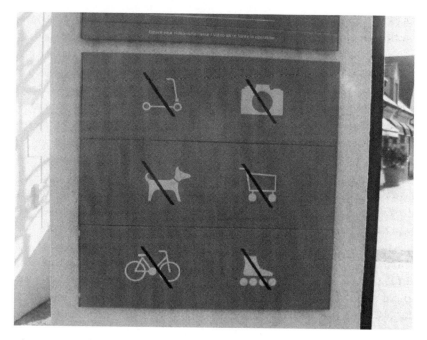

Plate 3.2 Regulating space: dos and don'ts at Marne la Valee shopping centre (photo: author)

series of playspaces but a linked series of (capitalist) functions – domestic reproduction, retailing, commerce, production, consumption, and so on. Play – in so far as it is planned for in the contemporary city – tends to be envisaged as something that only young children do, and then only in designated, safe playgrounds.

Consequently, skaters produce *spaces of representation* which oppose official *representations of space* and raise key questions about who has the right to the city (see also D. Mitchell 2003). In de Certeau's (1984) terminology, the skateboarders' use of the city constitutes a *tactical* appropriation. As he put it, the tactics of the weak do not 'obey the laws of place, for they are not defined or identified by it' (de Certeau 1984: 29); they are counterposed with the strategies of the strong. Moreover, such tactics may be self-consciously conceived as a form of opposition by groups which imagine themselves as part of the counterculture. This is reflected in the music, styles and fashions adopted by urban skaters, as Flusty (2000) suggests in his description of skaters in Los Angeles:

> From deep within the bowels of Bunker Hill, Pablo, Juan, Julio and Bob come rumbling out of the Third Street tunnel. Confronted by a red light and a river of one-way traffic at Hill Street, they kick up the noses of their skateboards in unison and, tails scraping against the sidewalk, come to a dead stop inches from the curb. Dressed in Chinos and baggy knee-length shorts, T-shirts and tanktops emblazoned with skate team and band logos, visored caps askew on their heads, these four comprise the core of Mad Dog Skate.
>
> (Flusty 2000: 153)

Like most skateboarders, the loosely affiliated skate team Flusty describes do not have the resources to build their own ramps and obstacles, so take to the streets, contributing to the evolution of a 'street style' of skating. Flusty (2000: 154) describes how they congregate around the Bunker Hill area – 'an agglomeration of low curbs, wide expanses of pavement, flights of gentle steps . . . long handicap access ramps for picking up speed, and retaining walls that protrude from plaza surfaces out into open air as the hill falls away beneath'. Although this an area providing opportunities for 'serious play', the skaters' occupation of this space is contested and often brings them into conflict with landowners and retailers. Skaters, according to local shop-owners, are 'noisy', 'disruptive' and engage in activities that endanger shoppers. Here, and elsewhere in Western cities, the police often caution skaters for 'pedestrian endangerment': infractions of skate bans in specific areas can merit fines, hours of community service and/or short sentences in juvenile detention facilities (see Woolley and Johns 2001 on the UK experience).

The idea that geographic order is imposed 'from above' through the panoptic gaze and segregationist strategies pursued by the police, magistrates, engineers and planners accordingly needs to be tempered with the observation that at 'street level' we find that individuals and groups create their own urban geographies, using cities in ways very different than bureaucrats and administrators intend(ed). Ideas of resistance are consequently of major importance in many geographers' discussions of the everyday politics of the city. Rejecting notions that power can be wielded only by the powerful, such work adopts a more diffuse notion of power (inherited from Foucault) to suggest that if the power to own and occupy space is everywhere then it can potentially be resisted everywhere. In such cases resistance involves attempts to create different spatialities from those that are defined by the law and state. This can result in places being changed or adapted as people become empowered through resistance (Sharp et al. 2000).

De Certeau's work is therefore highly significant given it draws attention to the everyday manifestations and forms of resistance that may be found in cities. Deflecting attention away from the more 'obvious' forms of resistance – squatting, sit-ins, strikes, riots, parades, to name just a few – de Certeau argued that resistances are more usually assembled from the materials and practices of everyday life. These resistances may therefore be ephemeral and slight; for instance, a single look, a movement or a simple spoken word might all encompass resistance. Developing this logic, de Certeau famously wrote of the simple act of walking as resistive, describing it as a 'process of appropriation of the topographical system on the part of the pedestrian' (de Certeau 1984: 97). Here, he implies that pedestrians can effectively reclaim the streets through improvisational tactics, with their footfall fleetingly appropriating spaces that have been rendered 'sterile' by engineers, planners and architects. The act of walking therefore becomes, in effect, one of the principal ways that citizens can refute the notions of moral and social order which have been inscribed on the landscape. Notable examples here include the way that many women's groups have been involved in attempts to 'Reclaim the Night' simply by walking on the streets of cities at night (Watson 1999); more overtly theorised instances include the situationist dérives which seek to expose the vicissitudes of the consumer city by enacting an unmeditated ramble through the city (see Box 3.2). Another theorised (and highly athletic) attempt to rewrite the city is to be found in the stylised activities of the free-running parkouristes who climb and leap across the cityscape in an often death-defying manner: their playful engagement with the city's hard landscape challenges all citizens to look at the city afresh (see Plate 3.3).

The representational mode of free-running – like skateboarding – is not interpretive, but performative. Overtly politicised street parties (such as those organised by Reclaim the Streets) and 'carnivals against capitalism' also celebrate the playful elements of urban life, staking a performative claim to space. Likewise, artistic interventions in the city – projections, street stencils, installations and 'billboard banditry' – are often playful in intent, seeking to subvert through strategies of détournement (Plate 3.4) (see especially Cresswell 1999). What all have in common is

Box 3.2 SITUATIONISM

Initially associated with a number of surrealist and avant-garde groups, the Situationist International (SI) formed in 1957 with the intention of re-enchanting everyday life. From the outset, the SI spoke to the conditions of contemporary urbanisation, and sought to expose the radical potential of urban space (noting that the contemporary city contained the conditions of possibility for its own transformation) (Pinder 2000). Inspired by the revolutionary actions of the French communards in the 1870s, the SI similarly sought to give the city 'back' to the people, adapting a number of playful concepts and tactics designed to emphasise the impoverished nature of everyday life in the city, Chief among these was the *dérive*, and experimental and unmediated drift through the city intended to show how the capitalist city held people in its thrall and condemned citizens to an essentially boring life. In turn, the *dérive* was intended to unfix the city's relations, and bring a more playful form of existence into being. The most celebrated proponent of the dérive – Guy Debord – saw the dérive as a way of tapping into the *psychogeographies* of place and understanding the way that cities are designed to evoke particular feelings or emotions. His own 'vagabond peregrinations' around Paris sought out disappearing spaces where malcontents and dreamers conversed, identifying points of resistance against an interminable process of 'improvement' (Merrifield 2005: 932). Romanticising those who lived an openly independent life, Debord and SI thus opposed the forces of modernisation which polished the rough edges of city life, and argued for a more 'real' (and less spectacular) city. During the student riots of 1968, situationist slogans (e.g. 'Beneath the pavement, the beach') were daubed on the streets of Paris, and for a while it seemed as though the urban revolution that the SI had sought was becoming a reality (Hollier 1994). The state's censure of some of the leaders of the 1968 revolt dissipated the influence of the movement, and threatened to relegate situationism to a mere footnote in art history. However, its ideas have had an enduring appeal for many musicians, artists, activists and academics. In particular, a spate of positive reappraisals in the 1990s (e.g. Plant 1992; Sadler 1998), coupled with the popularity of the psychogeographical writings of Iain Sinclair and Ian Home, means that situationism is currently receiving more attention from urban geographers than it did in the 1960s. As Pinder (2004: 118) argues, in an era where utopian thinking is badly needed, situationism provides a wealth of important ideas for those seeking to construct a critical geography which might attack the 'congealed qualities of the urban'.

Further reading: Sadler (1998); Hoggart (2004)

Plate 3.3 Le parkour: a free-runner turns the city into a playground (photo: Pauli Peura)

Plate 3.4 Stencil art in London: Banksy's urban interventions blur the boundaries of art, surrealism and situationism, and provoke a mixture of critical readings (with permission of the artist)

that they are *interruptions* which produce new urban temporalities, puncturing the humdrum routine of city life through ludic celebration. This principle of interruption is arguably that underpinning Breton's surrealism, Debord's *dérive* and Lefebvre's celebration of *la fête* (Roberts 1999). Each is based on the idea that everyday life contains moments that break through its linear repetitions to challenge processes of abstraction and ordering ('eruptions of instability through the carefully spread net of rational and homogeneous modernity' – Shields 1999: 183).

In their own ways, situationism, street stencilling, urban grafitti and *le parkour* are all deliberate attempts to subvert and subvent urban space. But if some politically motivated groups may set out to challenge codings of city space in a manner that deliberately flouts convention, others may find their presence in particular city spaces unintentionally brings them into conflict with the forces of law and order. As such, their *transgression* does not constitute an intended act of resistance but relies on that particular action being noticed and considered deviant and marginal (Cresswell 1996). Those who transgress are frequently subject to exclusion, being forcibly ejected or asked to leave specific places. An example here is the presence of street children in some Western, and many non-Western, cities. As Young (2003) relates in her study of street children in Uganda, for these children, survival may depend on 'street activities' such as picking vegetables in the markets as they 'fall' from the trucks, snatching wallets under the anonymity of a crowded area, or begging from passersby and the many shops and restaurants in the city. At busy traffic lights, children may smear dirty cloths across car windows, asking for payment, or else simply beg. Young (2003) suggests this type of begging activity often antagonises other street users because the children are very persistent: further, it is a reminder of the social inequalities that are evident in the city. A common response is for the state to introduce laws intended to remove these children from the streets (adapting a time-honoured logic of 'out of sight, out of mind').

More generally, homeless people constitute a group whose everyday spatial practice brings them into conflict with the authorities. As Daly (1998) points out, this group are forced to occupy public space for both economic and social reasons (e.g. bartering, begging, socialising, finding a place to live) because they have been evicted from the private spaces of the real estate market. However, their subsequent presence on the streets is fiercely contested, and although it is possible to detect a great deal of general sympathy for those living on the streets, at a local level this tends to translate into a concern that homeless people should be removed from particular residential and commercial areas. This urge to exclude the homeless has complex origins revolving around people's fears of difference, but certainly in societies where private property is highly valued, those without houses are often spurned as non-entities. The sight of homeless people using public space for activities normally restricted to the domestic realm (e.g. washing, urinating, sleeping) is one that appears particularly transgressive and disturbing to mainstream populations. The stigmatisation of homeless people thus resonates with the same kinds of metaphors that have been used to describe troublesome 'others' throughout history, with ideas of homeless individuals as dirty, deviant and dangerous provoking fears about the potential of homeless people to 'lower the tone' (and the house prices) in a particular area.

In an era when city governors are seeking to attract global investment through vigorous efforts at place promotion (see Chapter 2), this stigmatisation of homeless people is arguably becoming more pronounced. Targeted as a group who have no place in city centres catering for the affluent consumers, tourists and business-people, this is especially notable in Western city centres undergoing corporate gentrification. By way of example, Neil Smith (1996) graphically documents the way in which the 'improvement' and gentrification of New York's Lower East Side resulted in the systematic eviction of homeless people from a number of shelters, parks and streets. Most notably, he describes the forcible removal of homeless people from Tompkins Square Park, an area that had begun to be regularly used as a place to sleep by around fifty homeless people, recounting how the police waited until the coldest day of the year to evict the entire homeless population from the park, their belongings being hauled away by a queue of Sanitation Department garbage trucks (reiterating that notions of social purification rest on ideas that certain people are 'polluting'). He sees this heavy-handed treatment as symptomatic of the fears evident among the new middle and upper classes who had relocated to the Lower East Side; groups who regarded unemployed people, gays or immigrants as a potential threat to their dream of urban living.

Neil Smith (1996) describes such heavy-handed attempts to exclude the street homeless from particular sites as symptomatic of the *revanchism* (literally, 'revenge') evident in contemporary cities, where sometimes brutal attacks on 'others' have become cloaked in the language of public morality, neighbourhood security and 'family values'. According to him, manifestations of revanchism have become apparent throughout the urban West with the introduction of curfews, public order acts, by-laws and draconian policing designed to exclude certain 'others' from spaces claimed by the affluent. For example, in many US cities, this has involved the passing of 'anti-homeless' laws designed to prevent people congregating in parking lots, making it illegal to panhandle or beg within range of a cash machine or even making sleeping in public an offence (see D. Mitchell 1997). Elsewhere, suitable sites for the congregation of the street homeless have been 'designed out' by the authorities barricading public parks at night, removing benches that might be used for sleeping on. Collectively, these acts send out the message that the authorities wish to remove the homeless from the planned spaces of the consumer city (something underlined in the increasing espousal of Zero Tolerance policing for panhandlers, beggars, peddlers and prostitutes in many Western cities – see Hubbard 2004).

EVENTS TO COME: EMBODIED SPACES, EVERYDAY LIVES

Documenting the diverse forms of life that take shape on the streets clearly alerts us to the varied forms of transgression and resistance practiced in the city, and offers a useful rejoinder to official representations and mappings made 'from above'. As Cresswell (1996) details, focusing on the actions that are regarded as out of place in the city helps reveal the moral and social order immanent in everyday space (explaining the current popularity of studies of urban transgression). Yet dwelling on examples which disturb the established rhythms of the city potentially distracts us from considering the routine business of getting on and getting by in the city. After all, whether or not they are part of the homed or homeless population, all urban dwellers have to negotiate the city *practically*, and work through the dilemmas, problems and possibilities of 'getting by' in the city. An example is simply walking around the city, an everyday activity that is nonetheless dependent upon a range of purposeful and directed actions. Walking from one shop to another when we are shopping, for instance, rarely involves us walking a premeditated route but requires an adaptation to circumstances: we may be drawn along with crowds or cut through a back alley, cross roads at a pedestrian crossing or simply cross when we sense a gap in the traffic, stop to buy a copy of the Big Issue from a vendor or change direction to evade a petitioner. On the other hand, we may be that Big Issue seller, searching for a pitch that offers warmth, comfort and, it is hoped, a steady stream of potential purchasers. Whatever, the myriad of encounters we have to negotiate in the city requires a constant awareness and surveillance of other urban practitioners: the resulting 'street ballet' is the outcome of learnt modes of interaction designed to minimise contact with strangers and to deal with the complexity of the streets. Overall, the sum of our movements cannot be explained in rational terms (and is unlikely to be the shortest distance between a series of points).

Negotiating the everyday city, and connecting our home and work lives, thus involves considerable skill and aptitude. Jarvis et al. (2001) insist that much of this practical accomplishment remains underexamined by urban geographers. Notable exceptions are evident in the pioneering work of feminist geographers on social reproduction, which can be defined as concerning the world of non-work and unpaid work (i.e. activities such as shopping, housework, taking children to and from school). Perhaps more so than any other literature, it is this feminist research that has revealed just how much is missed by the planner, geographer or architect who remains fixated on the view from above. An example here is the effort required to navigate and negotiate the city when looking after children (not least those in pushchairs). Noting that childcare predominantly remains 'women's work', and, conversely, that most planners and architects are men, feminist research has highlighted the difficulty that can be faced when travelling from home to school, or from shops to home, when encumbered with children and heavy shopping. Widely cited studies such as that by Tivers (1985) reveal the tortuous journeys some childcarers have to undertake to avoid steep inclines or steps, as well as to avoid dark underpasses that may provoke fears of urban mismeetings and attack (see also Valentine 1989).

Such differential gender mobilities have been widely noted in feminist critiques of planning: for example, Greed (1994: 135) suggests that the division between drivers and pedestrians 'equates nicely with the division between shoppers and workers, and hence between women and men'. Supporting Lefebvre's (1995: 73) assertion that 'everyday life weighs heaviest on women', such writing suggests that modern planning often imagined a 'time-rich' and mobile housewife that, in practice, did not exist.

The fact that many comprehensive redevelopment schemes in post-Second World War British cities incorporated city centres replete with multilevel shopping centres, underpasses, and poorly lit and inaccessible multistorey car parks further underlines the dissonance between male planners' conceptions of the city and women's use and occupation of the city (Hubbard and Lilley 2004) (Plate 3.5).

Related perspectives on the (often-problematic) negotiation of urban space are provided in work on disabilism and the city, which demonstrates that cities do not cater for the full range of human body types and capabilities. From the perspective of a disabled body, the city is characterised by physical inaccessibility and exclusion, with the physical layout of cities at both a macro and micro scale placing disabled people at risk of both personal injury and social exclusion (Gleeson 2001). Hence, while the marginalisation of 'disabled' people is clearly connected to the privileging of certain bodies as productive, this is compounded by thoughtless

Plate 3.5 Post-war planning – exemplified here by Coventry city centre, UK – created multilevel precincts in which a maze of underpasses, overpasses, ramps and stairs created particular problems for parents and elderly people (see Hubbard and Lilley 2004)

urban design. For example, those with locomotion problems (a category which includes most, if not all, of us at some time of our lives) have typically found their access to spaces of work, leisure and welfare constrained by poorly designed buildings which prevent the use of wheelchairs or walking frames by including stairs or excessively steep ramps. The relative lack of accessible toilet facilities for wheelchair users further constrains their spatial mobility (Kitchin and Law 2001). Likewise, those with visual impairment experience a range of everyday mobility problems in the city, only some of which are mitigated by the installation of dropped kerbs, Braille signage and textured street paving.

Such examples indicate the sheer bodily effort, resourcefulness and adaptation which is needed to cope with urban life. Literatures on everyday practice thus point to the importance of *embodiment* in urban geography:

> One thing that does seem to be widely agreed is that place is involved with embodiment. It is difficult to think of places outside the body . . . think of a walk in the city and place consists not just of eye-making contact with other people or advertising signs and buildings but also the sound of traffic noise and conversation, the touch of ticket machines and hand rail, the smell of exhaust fumes and cooking food.
>
> (Thrift 2003b: 103)

Thrift's ruminations on city life serve to make the point that we cannot conceive of a city without thinking about the way it is experienced and registered via the body. Likewise, Edensor (2000: 76) insists that bodies act upon the city, 'inscribing their presence in a continual process of remaking'. And yet it is only since the late 1990s or thereabouts that geographers have truly engaged with debates about embodiment, initially via work on bodily ability and disability, but latterly in relation to a range of social identities and practices.

As has been suggested above, one way in which our everyday experience of cities varies is according to our sexed and gendered bodies (and other people's perceptions of our gendered bodies). Davidson and Bondi (2004) insist that women feel a heightened sense of embodiment, thanks in no small measure to the fact that men express themselves expansively, and impose masculinity in and on their environs. A common feeling for women is accordingly that they are caught and confined in city spaces.

One (extreme) example of this is the agoraphobic panic some women experience in social spaces and situations (Davidson 2000), yet this is perhaps just an exaggerated example of the way space presses differently on women's and men's bodies. This is also apparent in the exclusion of gay, lesbian and bisexually identified bodies from 'straight' spaces, with the majority of everyday spaces supporting a heterosexualised body ideal (Binnie and Valentine 1999). When bodies do not conform with this ideal, and cross established sex/gender boundaries (for example, through transvestism), they are often subject to intimidation and harassment. For instance, Browne (2004) describes how the traditional distinction of male and female toilets creates specific problems for those whose bodies do not conform to feminine or masculine ideals. For such reasons, some people may make major detours to avoid certain public toilets.

Such questions of the connections between gender, sexuality and the body are also highlighted in the work of Longhurst (2000). Drawing on feminist theorists including Luce Irigaray and Elizabeth Grosz, her work on the pregnant body provides an interesting example of how pressures to fulfil idealised body images make some women feel uncomfortable in certain urban spaces. Interviewing pregnant women in Hamilton, New Zealand, she found that the Centre Park shopping centre – a shopping environment targeting middle- and high-income women – was experienced as an exclusionary space. This sense of exclusion was evident on a number of levels; for example, it was evident that window displays often incorporated idealised images of women that were normatively glamorous, sexy and attractive, but which alienated pregnant women. Moreover, few shops provided clothing for pregnant mothers. More prosaically, those interviewed also reported that toilets, escalators and seating were not designed with the bodily form of pregnant women in mind. Concluding, Longhurst stresses there are many social norms which pregnant women are supposed to adhere to in terms of dress code, comportment and movement in public space. Because of cultural rules of pregnancy whereby women are not supposed to engage in sport, drink or smoke, or even have sex, it is therefore not surprising that some women found their usual public behaviours in public becoming unacceptable. In short, women whose bodies are sexed (and sexy) before they are pregnant take on different meanings during pregnancy, meaning that they may feel 'ugly and alien' in some city places that they previously felt welcome in. Sexist assumptions about women's rights to display their body may also serve to discourage women who consider themselves 'overweight' from sunbathing in public parks and beaches (McDowell 1999). Many 'obese' women also feel extremely uneasy about eating in public space for fear that their 'uncivilised eating' may be read as evidence of their gluttony (Valentine 1999).

This focus on the relations between the body and the city suggests that the social and biological entwine to shape and constrain urban routines. Likewise, it implies that geographers need to theorise urban settings as spaces that individuals engage with through both mind and body (social action being subjective, situated and embodied). However, Amin and Thrift (2002: 85) push this further, arguing the body remains the chief source of agency in the world, but that few bodily actions actually require motive (i.e. attribution of intention, justification or premeditation). For them, it is important to realise our material surroundings provoke bodily actions that, 95 per cent of the time at least, occur in the 'cognitive unconscious'. Simplifying to the extreme, this implies many social practices are intuitive – for example, we might cycle to work without remembering how we got from A to B, or play sports without thinking about what we are doing (or why). Action in space is therefore rarely conscious, and is often improvised. Amin and Thrift (2002) go so far as to suggest this improvisational bodily action includes talk: not just as representational praxis, but as a way of making sense of the world (through the process of making suppositions about our circumstances – see also Laurier 2001).

Amin and Thrift's (2002) privileging of embodied practice over mediated thought and language draws on the phenomenological theories of authors as varied as Mauss, Benjamin, Wittgenstein, Merleau-Ponty, Heidegger and Bourdieu (for an overview, see Hubbard et al. 2002). It also has some notable precedents in humanistic (i.e. human-centred) geography. For example, seeking to explore people's negotiation of the streets, David Seamon (1979: 16) developed a phenomenology of everyday life which sought to uncover and describe things and experiences – i.e. *phenomena* – 'as they are in their own terms'. Rejecting abstract theorisation and categorisation, he developed an experiential framework which focused on three related phenomena: movement (a focus on how bodies move through space), rest (how individual bodies find a place of dwelling) and encounter (how bodies interact with other bodies and things in their everyday worlds).

This 'triad of environmental experience' placed particular emphasis on the body as something which comes to 'know' its environment on its own terms. This may be an uncomfortable idea because we are so used to thinking about the ways in which our minds are our centres of experience, feeling and knowledge, but Seamon's interpretation suggested that our bodies too have intimate knowledge of the everyday spaces of our lives. The implication here is that we do not have to think about the way we move through urban space: our body feels its way. This idea is obviously opposed to the tendency of Western thought to separate the mind from the body, where the body is seen as largely under the control of the mind, as the tool of the mind (Butler 1999). Seamon is keen to break down this dualism. Instead of the human subject being theorised as a mind 'trapped' in, or working with, the body, he suggests that we consider individual people as 'body-subjects'. Following Merleau-Ponty (1962), the relation of a subject to the world is seen to be revealed in the purposive movements of the body – the body itself becomes the locus of intentionality (see also Cresswell 1999).

Despite the renewed interest in the phenomenology of the body, most geographical research remains predicated on the use of research methods that cannot adequately capture embodied experiences of space. The exceptions are informed by emerging debates in geography over the need to engage with lay knowledges (Crouch 2001) as well as the turn to non-representational theories (Box 3.3) that prioritise doing over discourse:

> The emphasis is on practices that cannot adequately be spoken of, that words cannot capture, that texts cannot convey – on forms of experience and movement that are not only or never cognitive. Instead of theoretically representing the world, 'non-representational theory' is concerned with the ways in which subjects know the world without knowing it, the 'inarticulate understanding' or 'practical intelligibility' of an 'unformulated practical grasp of the world.
>
> (Nash 2000: 655)

The privileging of 'ordinary' people's knowledge is crucial here, with the politics of non-representational theory stressing the importance of 'appreciating, and valorising, the skills and knowledges' of embodied beings that 'have been so consistently devalorised by contemplative forms of life, thus underlining that their stake in the world is just as great as the

Box 3.3 NON-REPRESENTATIONAL THEORY

For Thrift (1997: 146), non-representational theory is concerned not so much with representations and texts (see Chapter 2) but more with the 'mundane, everyday practices that shape the conduct of human beings towards others and themselves in particular sites'. This necessitates devoting attention to things that words and representations cannot express – the practical experiences of ordinary people that are rarely spoken of but constantly performed and felt. As he puts it, this focuses on the body, or more correctly, the body-subject, as it finds itself and remakes itself in a world that is constantly changing and becoming. Exploring the non-verbal and pre-discursive ways people 'do' identities, there is now an extensive geographical literature on the ways people dance, walk, swim, ride or sit their identities (Nash 2000), as well as the way they perform their identities through bodily gestures, modes of eating or styles of dressing (Valentine 1999). This literature has illuminated our understanding of an ever-widening range of everyday and seemingly mundane behaviours, such as cinema-going (Hubbard 2002), gardening (Crouch 2001), listening to music (B. Anderson 2004), drinking (Latham and McCormack 2004) and driving (Merriman 2004; Thrift 2004; Laurier et al. 2005). In relation to driving, for example, Thrift (2004: 58) suggests it is 'possible to write of a rich phenomenology of automobility, one often filled to bursting with embodied cues and gestures which work over many communicative registers and which cannot be reduced simply to cultural codes.' However, critics of non-representational theory (and there are many) assert it offers little new in terms of theoretical explanation. Moreover, some are dismayed that it proclaims to be anti-representation. Yet those advocating non-representational approaches insist they are not anti-representation, and that the labelling of this approach is more than a little misleading. Lorimer (2005) surmises it is better to talk of 'more-than-representational' geography, the original 'non-' title discouraging some from engaging with this important body of work.

Further reading: Thrift (2003b)

stake of those who are paid to comment upon it' (Thrift 1997: 126). Or, as Laurier (2001) has put it, it is about valuing people's everyday competencies rather than the world-views of theory-driven, professional researchers.

This shift from *theory* to *practice* is decisive as it demands that researchers consider urban spaces as embodied and lived, not just imagined and represented. Here it is interesting to note that one of the inspirations listed by Nigel Thrift (1999) in his tentative proposal for non-representational theories is Michel de Certeau. Thrift especially seizes upon de Certeau's description of the urban body as subject to social controls but endlessly expressive and resistive. Thrift (2003b: 109) echoes this when he argues that people are able to use talk, gesture and bodily movement to 'open up pockets of interaction over which they have control'. Malbon's (1999) ethnography of urban nightlife amply illustrates this, demonstrating that bodily practices and proficiencies (dancing, adornment, poise and so on) create choreographies of belongingness which often escape rational explanation or social control. Whether on dance floors, ice rinks or in crowded bars, people habitually use their bodies to make spaces of interaction, creating particular forms of conviviality and sociality. Of course, it may also be that the effects of alcohol and drugs are implicated in the particular ways of moving, gesturing, walking and talking that are in play here (Latham and McCormack 2004), while music is especially crucial to spaces of clubbing, creating particular moods through its rhythms (see also B. Anderson 2004). This points to the role of various *mediators* in the unfolding relationship between the body and the city – and hence a need to take the city's materialities seriously.

BEYOND REPRESENTATION? THE MATERIALITY OF CITIES

Traditionally, geographers have not ignored the materiality of the city, and have sought to map and model the city's built landscapes in a variety of ways (see Chapter 1). Yet a heightened responsiveness to matter and materiality has arguably emerged from the 'anxiety about the position of "the material" in the twists of the "cultural turn"' (B. Anderson and Tolia-Kelly 2004: 672). Most significant here is the argument that the recent preoccupation with representation has distracted from consideration of the non-representational. For example, Loretta Lees argues that to analyse the power and ideology encoded in the built environment is not enough: 'architectural geography should be about more than just representation' (Lees 2001: 51). Lees consequently argues that architecture is performative 'in the sense that it involves ongoing social practices through which space is continually shaped and inhabited' (Lees 2001: 53), and therefore calls for a 'critical geography of architecture' which would explore the use and occupation of everyday spaces. For Lees, it is important to move from questions of production to questions of consumption, detailing how urban landscapes 'work' within cultural practice (see also Llewellyn 2003).

Returning to Matless's (2000) warning about neglecting the materiality of landscape, it is here we start to realise the limits of the textual metaphor; landscape is not simply perceived, it is used, occupied and transformed. Similarly, it is not only read, but also smelt, heard, felt and lived. In Lees's own studies, such questions of habitation are addressed through ethnographies of urban spaces designed to elucidate not only what buildings mean, but also what they do (in the sense of what dominant social relations occur in and around them) (see Lees 2001, 2003). Elsewhere a desire to chart how buildings are used has encouraged geographers to use interview methods to allow people to explain how they use and interact with their surroundings. By giving voice to residents and inhabitants of architectural spaces, geographers not only develop a 'polyvocal' narrative, but also highlight the way projected meanings of the city are subverted and changed through occupation. For example, Nick Fyfe (1996) has used poems to explore the contestation of the Clyde Valley plan in the 1950s, suggesting that poetic visions of Glasgow written by local residents offer a 'thick interpretation' of urban change, richly complex and contradictory. In other studies, autobiography, written testimonies and letters to local newspapers have been studied to explore the way space is occupied and lived (Finnegan 1998). A further example of this is the anthropological account offered by Holston (1991), who uses interview, archival and ethnographic evidence to highlight how the planners' conception of Brasilia was undermined by residents as they sought to recapture the atmosphere of traditional Brazilian street life and (literally) turned their backs on the pedestrianised precincts intended to act as neighbourhood centres.

Considering geographers' gradual attunement to the materiality of the city, Latham and McCormack (2004) are nonetheless sceptical about the separation made (by Lees and others) between the material – described variously as the actual, the concrete and the real – and the immaterial, the abstract and the unreal (see also Pile 2005 on what constitutes 'real cities'). For Latham and McCormack, the problem with 'culturally inflected' urban geography is 'not the fact that it has engaged excessively with the immaterial, but, in contrast, that it has not engaged with sufficient conceptual complexity with the importance of excess to any notion of the material' (Latham and McCormack 2004: 704). Their argument is that the things and objects which might be commonly regarded as material are no more or less real than the effects of language and discourse: both involve relations of variable duration and force. This allies them with Marcus Doel's (2004) conception of a post-structural geography which is not obsessed with language at the expense of the material, but is interested in how language enters into the constitution of the world:

> Contrary to popular opinion, we [post-structuralists] do not wish to elude the gravitational pull of the world in order to float freely among signs and images. Rather, we affirm the falling back of signs and images into the play of the world. We remain – as always – resolutely *materialist*. So, we are struck by the *force* of signs, by the *intensity* of images, and by the *affects* of language . . . we no longer recognise anything other than material and immaterial forces, the differential relations between forces, and their incessant shuffling. Whatever there may be, it always *strikes* someone or other as an articulation of force.
>
> (Doel 2004: 150–151, original emphases)

The implication here is that signs and symbols cannot be defined in opposition to some material, concrete notion of the city as bricks and mortar. Or, to put it another way, language has a life of its own, every bit as real as that which we might regard as material. Hence, Latham and McCormack (2004) contend that the materiality of the city *emerges* from the relations of different bodies, machines, words, images and signs – none of which can be assumed to be more real than any other. Materialising the city evidently involves more than just considering its objects.

The idea that the city is the urban is an ongoing outcome of the interaction between a myriad of 'small-scale self-organizing processes' suggests the need for theories of the urban that do not search for an overarching logic but instead attend to the ways that the remarkable plurality of substances and relationships give reality and shape to urban life (Latham and McCormack 2004: 709–719). One of the examples which Latham and McCormack use to illustrate their conception of urban materiality is the relationship between car and driver – one which is woven into the city in a number of ways. One need only think of the proliferation of car parks, parking meters, drive-in cinemas, drive-through restaurants, service stations, lay-bys, roundabouts, flyovers and underpasses to realise the profound impacts that cars have had on urban life. As Urry (2000: 59) details, 'the car's significance is that it reconfigures civil society involving distinct ways of dwelling, travelling and socialising in and through an automobilised time-space'. Surprisingly, however, there have been few accounts which have considered the complex ways that the car has been incorporated into everyday urban life. Furthermore, geographers have only recently begun to document the side-effects created by the dominance of the motor car, such as the aggressive geopolitics of oil production, the rise of environmental pollution, and a litany of traffic accidents (see T. Hall 2003).

Clearly, the motor car is involved in the making of cities in a number of ways. Yet Thrift (2004) insists that the car has been largely analysed in purely representational terms by cultural commentators, for instance, as the symbolic manifestation of various desires or dreads. Less frequently, the emotional relationship which the driver has with their car is considered, with the problematic ability of the car to act as extension of Self discussed (Lupton 1999). In contrast, car culture has rarely been written of as an emergent set of machinic relations between practices and technologies, involving, for example, the development of highway rules and regulations, modes of driverly conduct, road signs and markings, street lighting, radio travel warnings, road atlases and satellite navigation systems. This 'complex heterogeneity' of technical machines, petrochemical machines and corporal machines is constituted not only in the design and regulation of road spaces, but also in the materiality of the automobile itself, which makes 'particular assumptions about the proper relationship between car and driver, driver and passenger, car-human hybrid and car-human hybrid, car-human hybrid and the road surface, and on and on' (Latham and McCormack 2004: 712).

Hence, although the 'normalised and individualised' figure of the driver, and the 'mass-produced yet invariably customised vehicle' would appear to be the central elements of urban car cultures, Merriman (2004: 159) suggests it is futile to attempt to understand the 'performances, movements, semiotics, emotions and ontologies' associated with driving by attempting to distinguish between these. Further, a focus on materiality suggests the intimate relation between car and driver represents only one aspect of the material cultures of driving:

> While academics have explored the more durable or familiar practices and relations of hybridised car-drivers . . . many other 'things' became bound into the contingent and momentary orderings of these hybrid figures. Legislative codes, roadside trees, service areas, 'cat's eyes', passengers, wing mirrors, cups of tea, tarmac, fog, and various experts may serve as constituent elements in the relational performance of . . . driving.
>
> (Merriman 2004: 162)

In his own studies of UK motorway driving, Merriman (2005) thus details the instruments associated with engineering, the judiciary and government (including road engineers and road safety experts) designed to affect the performances and movements of drivers and vehicles and to ensure safe and commodious travel. Experimental crash barriers, fog warning signs, anti-dazzle fences/plants, speed limits and new driving codes all affected the performance of motorway driving, not just as mediators, but as constitutive elements in the material spaces of the motorway.

Work such as Merriman's serves to demonstrate that a focus on the phenomenologies of everyday life involves an appreciation of the multiple relations between subjects and objects, and need not centre on the (limited) agency of the human body-subject. The realisation that the agency of the city is widely dispersed, and the cities are more-than-human, is accordingly an important reminder that the materiality of the city needs to be taken seriously indeed. Concurring with this view, Latham and McCormack (2004) conclude that, for all its achievements, traditional urban geography has presented us with a remarkably emaciated view of what cities consist of. Extending our purview of urban process to consider the remarkable plurality of substances and relationships that give reality and shape to urban life thus represents an attempt to do justice to the materialities of the city. Of course, questions remain about which of these relations and substances should most concern us as human geographers. This is something we will explore in Chapter 4, where I consider some other ways in which geographers are attempting to theorise the non-human, more-than-human and even posthuman aspects of city life.

CONCLUSION

Through its engagement with cultural and social theory, urban geography has taken on board a range of ideas about the importance of urban representations and meanings. While these undoubtedly enter into the constitution of everyday life in a variety of ways, many urban geographers have regarded the cultural turn as steering the discipline into an increasingly arid landscape of linguistic deconstruction, far removed from the creative practices and routines of city life. No matter how misplaced such criticisms of geography's linguistic and representational focus have been, a common tendency has been to assert the need for non-representational accounts of urban life. In essence, non-representational theories contend that we cannot hope to comprehend the use and occupation of urban space solely by exploring the social meanings projected onto the urban landscape: additionally, we need to explore how these settings are embodied through performance. After all, cities may be scripted, but our performances do not always follow the script. This necessitates an exploration of the sensuous and poetic dimensions of embodiment – an endeavour that may require the development of new research methods given the limits of standardised research techniques (particularly questionnaires) for elucidating pleasures and pains that may escape rational explanation. Yet at the same time, it is apparent that this requires consideration of the way practices are negotiated in relation to representations, given that discourses imbue practices with meanings that are both spoken and felt.

This chapter has thus charted a rich seam of work exploring the practical and embodied skills which are woven, almost subconsciously, into the lives of urban citizens. Although some of this work appears guilty of romanticising the banal and repetitive aspects of urban life, and endowing them with a revolutionary potential that is rarely realised, collectively it signposts some important directions for urban research. Summarising this potential Lorimer (2005: 85) seeks to distil the essence of non-representational theory, suggesting 'it is multifarious, open encounters in the realm of practice that matter most'. As he describes, this implies a need to think through 'locally formative interventions in the world', including phenomena that may seem remarkable only in terms of their apparent insignificance:

> The focus falls on how life takes shape and gains expression in shared experiences, everyday routines, fleeting encounters, embodied movements, pre-cognitive triggers, practical skills, affective intensities, enduring urges, unexceptional interactions and sensuous dispositions. Attention to these kinds of expression . . . offers an escape from the established academic habit of striving to uncover meanings and values that apparently await our discovery, interpretation, judgement and ultimate representation. In short, so much ordinary action gives no advance notice of what it will become.
>
> (Lorimer 2005: 92)

This emphasis on the immanent and the *becomingness* of the world is suggestive of an approach to human geography which is open to the world. This, then, is assuredly not an attempt to develop a 'grand theory', but an attempt to engage with the complexity of existence, charting out the lines of force and effect which give the city substance and meaning. In the following chapters, I consider some of these forces in more detail, and also explore how we can make sense of these in a world that appears increasingly complex and connected.

FURTHER READING

The increasing attention devoted to street geographies is nicely summarised in the various chapters of Nick Fyfe's (ed.) (1998) *Images of the Streets*: chapters discuss the geographies of the homeless, street children, disability on the street and street consumption. Amin and Thrift's (2002) *Cities* represents an important intervention in the literature on urban life, and is (partly) an attempt to demonstrate the implications of non-representational theory for urban studies. Robyn Longhurst's (2001) *Bodyspace* provides an effective summary of the relation of the body and the city, with case studies of domestic, leisure and business spaces demonstrating that the expressivity of the body is subject to control on a variety of scales. Richard Sennett's (1994) *Flesh and Stone* is a remarkable and expansive volume that considers the unfolding relationship between human physicality and the city. Finally, it is worth noting that geographers' prosaic attempts to elucidate everyday geographies often pale into insignificance when compared with the poetic street-writings of psychogeographers, flâneurs and journalists: the writing of Iain Sinclair is always worth dipping into for a different take on the city (see especially *London Orbital* 2003), while the contemporary 'fanzine' *Smoke* contains an entertaining mixture of photo essays and paeans to London. Online blogs (i.e. web-logs) of city life are also a fecund source of writing about everyday journeys and the spaces of the city: try especially http://london bloggers.iamcal.com for some distinctively ordinary takes on city life!

4

THE HYBRID CITY

Future researchers will take as given something we can only dimly perceive
today – and then may be too horrified to admit . . . Namely, that all
performance is electronic, that the global explosion of performance
coincides with precisely the digitalization of discourses and practices,
and that this coincidence is anything but coincidental.

(McKenzie 2001: 29)

It often seems the contemporary imagination is haunted by fears of
technology. The rise of mobile technologies, the way we work online,
new forms of data logging, satellite navigation, cybernetics and wearables
all suggest that the boundary between people and technology is dis-
solving (Schilling 2005). Likewise, the idea that new technologies are
driving cities towards a post-human (and possibly apocalyptic) landscape
of surveillance and discipline is a popular trope in city writing – one
that finds sharp expression in the contemporary *noir* urbanism of Mike
Davis (1990) or Paul Virilio's (2005) account of the *City of Panic*. Yet it
is important to note that the significance of new technologies may
be no more profound than those which preceded them. A brief trawl
through the histories of new technologies – printing, steam power,
gas lighting, electrification, the motor car, the telephone, television,
computers – confirms that all innovations are greeted with some level of
moral approbation and talk about the erosion of society. Furthermore,

each had impacts on the shape and form of cities that were, in their time, every bit as profound as is the perceived impact of cybertechnologies today.

The tendency to talk of new technologies in hyperbolic terms is unfortunate. One consequence is the relative neglect of 'past' technologies, and a failure to think critically about the new social formations that are associated with successive technological innovations. In geographical circles, this neglect is related to recent theoretical tendencies within the discipline. As discussed in previous chapters, while the 'representational turn' was broadly welcomed as a way of widening the ontological remit of urban studies, some have alleged that its focus on the immaterial and imaginary distracts from the 'concrete realities' of city life. For example, Kaika and Swyngedouw (2000) argue that technologies of flow have been strangely absent in many recent accounts of urban life – even though we are constantly surrounded by water pipes, electric cables or computer networks which hum, gurgle and pulse as we go about our urban lives.

While some of the criticisms of geography's representational turn are perhaps misplaced (after all, imaginaries have a materiality), there is little question that non-representational theories have rejuvenated interest in the multiple technologies and materialities of city life. In this chapter, I therefore consider some of these materialities, exploring how urban theorists have tried to make sense of science and technology as makers of the urban landscape. Here, I describe a move from accounts where technology is allocated a determining role in urban life to those where it is regarded as entwined in a more complex process of city-making. In so doing, we will begin to see that many current ideas about the place of technology in cities actually problematise our definitions of human agency, and the extent to which cities are made by people. Alighting on the concept of hybridity, I thus conclude by overviewing the polymorphous materiality of the city, alerting to the multiple forms of life which abound in the urban realm. In so doing, I explore how contemporary urban scholarship is currently trying to take account of the different agents (or actants) that animate our cities, be they people, machines, plants or animals.

THEORISING CITY INFRASTRUCTURES

As was noted in Chapter 1, the distinction between 'rural' and 'urban' ways of life has often been theorised in terms of progress and technological sophistication. To simplify: it has been frequently asserted that the city is a site of innovation in science, medicine, transport, engineering and building construction (Mumford 1961). In contrast, the rural is seen as technologically backward, with many new inventions and technologies making their presence felt in the countryside only years after they have become widespread in towns and cities. Further, some have argued that technology provides the *raison d'être* for the formation of cities. Notably, most theories intended to account for the emergence of cities have allocated a pivotal role to technology. For instance, many argue that urban centres first emerged as agriculture began to create a surplus of goods that needed to be stored, managed and redistributed. Yet for such surpluses to be produced, irrigation was essential. Identifying the emergence of the *hydraulic* society, Wittfogel (1957) stressed that large-scale waterworks were essential in the creation of surplus, with the control of floods by elaborate systems of dams and channels engendering both increased production and concentrated population settlement.

In Wittfogel's account, particular technologies 'discovered' in the third and second millennia BC (such as the use of underground Karez wells and stone-built canals) increased the food supply, allowing larger numbers of people to agglomerate into towns and cities. Simultaneously, because farmers were vulnerable to attack, armies were needed, with the implication of an officer class. Specialisation of labour brought the emergence of potters, weavers, metalworkers, scribes, lawyers and physicians, while the new surpluses also created the basis for commerce. This more complex economy required records, so writing (of which the first examples come from the bookkeeping records of the storehouses in ancient Mesopotamia) was born. As Wittfogel (1957) spelt out, the centralisation of administrative and political power among a hydraulic elite – who effectively wielded the power of life and death through control of water – further encouraged concentrated settlement.

Though the archaeological support for Wittfogel's thesis is mixed, his theory provides a good example of an often-told story of the links between cities and civilisation. In this story, the catalyst to the formation of cities is inevitably the discovery or application of a new technology quickly seen to have profound effects for the organisation and reproduction of society. For Wittfogel, irrigation technologies were crucial; elsewhere, other technologies are deemed to have played this pivotal role, with Childe (1936) famously identifying the plough, the invention of the wheeled cart and draft animals, sailing ships, the smelting of copper and bronze, a solar calendar, writing, standards of measurement, irrigation ditches, new craft technology and innovative methods of building as all implicated in the 'urban revolution'.

While most now concede that the search for urban origins is a great deal more complex than Childe (1936) or Wittfogel (1957) suggest (given that the emergence of a religious or military elite may have been as important in many contexts as the emergence of a technocratic elite), connections between technology and urbanisation have been stressed in many histories of urban change. The identification of cities as being either pre-industrial, industrial or post-industrial is one that is frequently made on the basis of the technologies that emerged as cities took on new forms and functions. For instance, many of the prime centres of British manufacturing in the nineteenth century (Birmingham, Manchester, Glasgow) exploded in population terms as new innovations (textile machining, steam engines, iron and steel metallurgy) enabled entrepreneurs to effect an 'Industrial Revolution' which required pools of pliant labour and a ready market for industrial commodities. The movement of bulk goods by canals and then railways further helped concentrate people in factories and the housing congregating around them. These infrastructure networks unevenly bound city spaces together, bequeathing new material divides as they centralised some communities and displaced others (see also Harvey 1985).

Other, less celebrated, infrastructure networks – water, sewage, gas and, ultimately, electricity – were implicated in the making of ever-more populous industrial cities. Each of these had multiple impacts on urban forms and practices. For instance, it was not until the nineteenth century that extensive networks of gas piping allowed for the effective lighting of street spaces. Gas-making was a costly and dangerous business, and it was not until municipal authorities began to invest in major gas works that such lighting became commonplace in the largest towns and cities. This first occurred in Britain, and, more particularly, in London, where the London and Westminster Chartered Gas Light Company was founded in 1810. By 1823 there were 215 miles of gas-lit streets, with 40,000 public gas lamps and, by 1852, 360,000 street lights. In contrast, rural areas did not generally have gas at all until the latter part of the nineteenth century (Nead 2000). Such 'industrial' gaslighting had multiple impacts: it allowed for the safe passage of workers to and from factories at all hours; it extended the hours of leisure (creating the possibility for urban nightlife); it allowed the eyes of the state and law to peer into the darkest nooks and crannies of the city. Moreover, it became a spectacle in itself, with new illuminated street spaces becoming a magnet for promenaders, who would marvel at the intensity of the light (Schlör 1998). Moreover, street lighting became a visible symbol of modernisation (and an obvious target for anti-state seditionaries, who often indulged in lantern-smashing) (Schivelbusch 1988). In its turn, electricity was to have similarly far-reaching impacts on urban life. Nye (1990) suggests that electrification represents a set of technical possibilities that have been selectively adopted by city governors to create and market particular packaged landscapes – for instance, the streetcar suburb, the amusement park, the 'Great White Way', the assembly line, the electrified home, the department store and so on (see Plate 4.1). Nye (1990) thus details how electricity came to touch every part of American urban life, taking centre stage in modern cities forged around an ideal of high energy and speed.

Inevitably, other technologies were associated with the emergence of industrial cities, not least those that measured industrial time. Historians have often maintained that industrialisation involved the replacement of traditional notions of time by modernised time (Kern 1983). Crucial here was the shift from qualitative to quantitative time, wherein agricultural senses of time shaped by diurnal rhythms of daylight and darkness (and the changing of the seasons) gave way to the continuous, precise time of the clock. This is frequently associated with transitions in the nature of production, with industrialisation replacing task-oriented work with time-oriented work (and a Puritanical work ethic that abhors 'wasted' time). However, the dominance of clock time has often been resisted, with the continuation of sensed and felt time attesting to the contested nature of time in everyday life. Corbin's (1998) exploration of the role of village bells in nineteenth-century France brings these issues into sharp focus, detailing how, at a time when public and private clocks were rare, and watches owned only by the elite, the sound of church bells marking services or ceremonies provided a means for villagers to become attuned to both their own biological rhythms as well as the seasonal and diurnal cycles of 'nature'. When the church began to mark clock time through the

Plate 4.1 *Scientific American* April 1881 celebrating Edison's achievement of providing electrically powered street lighting in New York (with permission of *Scientific American*)

same village bells – often under duress – the contrast of remorseless clock time and cyclical (sacred) time became acutely apparent: indeed, the French peasantry often protested furiously about the apparent discrepancy between 'their' time and the 'cosmic time' that was introduced by the French state in 1891 (and which was readily adopted in the cities). Irrespective of such protests, Corbin (1998: 112) concludes that nineteenth-century bell-ringing responded to the 'ever-more clamorous demands of modernity that were driven by the need to come to terms with all-embracing systems for the measurement and evaluation of time'. Given that cities increasingly moved to the rhythm of the capital (and not the cycles of nature), the logic of clock time was also manifest in urban workplaces, where it played a central role in the surveillance and appraisal of workers (who were forced to 'punch the clock' on entering and leaving work).

Corbin's (1998) remarkable account suggests the replacement of qualitative, sacred time by the mechanised rhythms of clock time amounted to a 'revolution in the culture of the senses'. Schivelbusch (1988) likewise speaks of modernisation as representing the industrialisation of experience, with technological innovations such as street lighting having profound impacts on what Thrift (1996a) terms 'time consciousness'. Accordingly, new devices were required to make sense of these experiences: for example, with church bells complemented, and ultimately superseded, by the pocket watch, the public clock, the timetable and other technologies that quantified experience in hours, minutes and seconds. New media – the photograph, the telegraph, film, newspapers – were also representative of this timed modernity. Embedded in the city, these media offered novel ways of recording and circulating modern time, simultaneously helping citizens adjust to the new rhythms of modern urban life (Amin and Thrift 2002). The fact many of these devices originated in cities has been interpreted as significant. For instance, Donald (1999) argues that early cinema should be understood as an urban phenomenon not merely because cinemas were located in the city, but because they were of the city, using new narrative devices, visual technologies and editing techniques to communicate the pace of modern urban life.

While writing on the changing experiences of time has accordingly begun to highlight the distinctive choreographies of industrial cities, it has often perpetuated a rather simplistic opposition between fast urban spaces and languid rural ones. This is evident in any number of lists of the symptoms of industrial cities: for example, the increasing quantity of traffic, the rapidity of financial transaction, the remorseless tempo of the assembly line, the telescoping of communication networks and so on (Highmore 2002a). Depictions of the incessant *speeding up* of society consequently abound in accounts of nineteenth-century industrial cities, implicated in the 'annihilation of space by time' that allowed all aspects of capitalist life to be extended and intensified. For example, Harvey's (1982) celebrated exegesis of Second Empire Paris revolves around the idea that Haussmann's new boulevards allowed for a more rapid circulation of commodities. Claiming that the boulevards were corridors of homage to the power of money and commodities, Harvey began to outline the negative consequences of the increased pace of urban life for Parisians, hypothesising a crucial link between modernisation and the lubrication of 'circuits' of capital. Hausmannisation may have brought a fantastic and elusive quality to life in the city but, from the perspective of the pedestrian, rationalising the distribution of capitalist products transformed the entire urban scene into a 'moving chaos' as the tempo of modern city traffic 'imposed itself on everybody's time' (Berman 1983: 159). In cities orchestrated by the flow of commodities, the *quantity* of human encounter may have therefore increased, but the *quality* of those interactions as the streets became devoted to commercial transaction and the associated circulation of goods and capital (Benjamin 1999).

The dehumanising effects of nineteenth-century Haussmannisation have thus been held up as exemplary of the deleterious effects of technology and speed on social life: however, Robins and Webster (1988: 49) have argued that it was only with the advent of Fordist assembly-line production in the twentieth century that time was required to be used 'as intensively, deeply and productively as possible'. This necessitated the strict demarcation of work/personal time and a regimentation of the movement of goods and people in the city. Both of these were deemed dependent on centralised transport planning and the *engineering* of circulation (hence, Le Corbusier's metaphor of city as *machine*). Berman (1983) suggests this high-modernist obsession with generalised mobility was to reach its apogee in Robert Moses' New York, where inner-city 'villages' were bulldozed to make way for freeways:

This new order integrated the whole nation into a unified flow whose lifeblood was the automobile. It conceived of cities principally as obstructions to the flow of traffic . . . as junkyards of substandard housing and decaying neighborhoods from which Americans should be given every chance to escape.

(Berman 1983: 307)

Moses' redevelopment was on a huge scale, yet the construction of vast urban highways linking the city to the country was a recurring strategy underpinning high modernity – as can be seen in the development of the Seine expressway in Paris (Ross 1994), or the urban motorways that were to connect London, Glasgow and Birmingham to a national motorway network in the United Kingdom (Fyfe 1996; Gold 1998). In each case, the destruction of established communities seemed a small price to pay for guaranteed mobility (especially when the sacrificed areas were working-class districts depicted as in urgent 'need' of modernisation). In other cases, new settlements and towns were planned from first principles around the motor car – the state capital Brasilia providing a prime example of how planners sought to orientate city life around the automobile. As Holston (1991) demonstrates, this was to prove less than successful in many respects, with the new inhabitants seeking to recapture the buzz of traditional Brazilian public life which they felt was lacking in a city of single-purpose highways rather than multifunctional streets, reorienting their businesses and homes away from the highway and embracing a slower pace of life.

While projects of infrastructure and transport improvement may be resisted, according to the French theorist Paul Virilio, those who ultimately control channels of movement in the city constitute a powerful urban elite. Latterly, however, physical channels of movement in the city – along which people, goods and traffic can flow – have been increasingly supplemented by virtual connections (Box 4.1). Enabled by successive innovations in data storage, transmission and processing (e.g. fax technology, mobile telephony, intranet and Internet connections, and wireless computing), the years since the mid-1970s have often been described as a period of 'communication revolution', allowing us to arrive at a point where virtual movement has seemingly produced a new 'informational society' based on the circulation of ideas and knowledges (leading to talk of weightless economies). As such, Robins (1995) identifies the virtual city as the logical outgrowth of projects of urban modernisation and technological improvement, replacing the centrally planned

Box 4.1 VIRTUAL CITIES

Much has been written about how urban work and play are transforming as more and more interactions occur in the virtual realm rather than the physical spaces of the city. It is the French urbanist Paul Virilio who has arguably done most to encourage geographers to think through the implications of this. Above all else, Virilio is known as a theorist of speed who has documented the ways in which life is speeding up. In this sense, Virilio offers a variant on the idea that time is annihilating space (see also Chapter 5), although his description of speed politics adds an important new slant. Although some have (rightly) suggested that Virilio overstates the extent to which chronopolitics (a politics of time) has displaced geopolitics (a politics of space) in contemporary society, his discussions of the creation of the *polis* provide a nuanced take on the imbrication of space and time (Luke and O'Tuathail 2000). For example, in *Speed and Politics*, Virilio (1986a) argues that cities are dwelling places organised through a fixed territorial infrastructure (of road and rail networks, ports, airports, etc.) that permits bodies to move. As he stresses, each of these has its own speed limits and regulations. The city itself is a conglomeration of these channels, a system of 'habitable circulation' flowing through an urban environment that is otherwise fixed and immobile (Virilio 1986a: 6). For Virilio, urban life unfolds in these channels with their movements, institutions and events. Virilio suggests that a politics of speed is thus about regulating the passage of people (and goods) through these channels, creating a position in which some flows are expedited, others downgraded. However, Virilio goes on to argue that recent developments in the fields of telecommunications have led to an erosion of the physical to the point where space is no longer a resource, but a burden. The implication is that in the era of the Internet, cyberspace and the global mass media, physical boundaries have been effectively overcome by systems of electronic communication and surveillance. For instance, the computer screen effectively allows us to access a world of information – or, to put it another way, to travel without moving (hence, Virilio's talk of an *inertia society*). The implication here is that physical meeting spaces are increasingly redundant, and technologies such as the automobile obsolescent. Virilio consequently proposes that 'a new kind of distance, the light (zero sign) of the new physics, suddenly takes the place of the customary distances of time (positive sign) and space (negative sign)' (Virilio 1999: 58). Widely read across the social sciences, Virilio has nevertheless been criticised for exaggerating the extent to which technology annihilates space, given that the impacts of speed are much more variegated than his general analyses imply. Likewise, his somewhat 'relentless negativism' about technology reifies the myth that new technologies destroy society – whereas, in truth, they bring new social formations into being (Thrift 2005b: 338).

Further reading: Virilio (1991); Doel and Clarke (2004)

transport networks characteristic of industrial cities with flexible (and individualised) informational networks. As the work of Virilio stresses, in the inertia society, people can travel without moving. At the same time, virtualism allows for new forms of living and working, with William Mitchell (2004) speculating that wireless laptop technology means more important work is performed in 'unassigned spaces' – hotel rooms, coffee shops, airport departure lounges – than in formal office spaces.

As we have seen, it is easy to get swept up in the hyperbole surrounding virtualism. But we should remember that virtual flows still require the construction and maintenance of complex physical infrastructures in the form of dense webs of optic cables, broadband telecom connections and local area networks (LANs). Moreover, dedicated servers and data storage devices need constant maintenance and management. Manuel Castells' celebrated treatise on the information age thus points out that while virtual technologies allow for remarkably distanciated transaction, they require spaces dedicated to the maintenance of this information superhighway. In Castells' (1996: 412) summation, this means the world of places – consisting of bounded and meaningful places such as the home, city or nation state – is being superseded by 'non-places' characterised by 'purposeful, repetitive, programmable sequences of exchange and interaction': examples include the placeless spaces of call centres, computer helpdesks, corporate offices and cybercafés where flows of information are fussed over, interpreted and transmitted. But it is not just space that is unravelling as virtualism becomes the norm. For Castells, real virtuality also alters the formal qualities of time. Within communication and media technologies, he contends that temporality is erased, suspended, and transformed. It is a 'timeless landscape of computer networks and electronic media, where all expressions are either instantaneous or without predicable sequencing' (Castells 1998a: 350). Castells' (2001) *Internet Galaxy* accordingly describes a realm of *timeless time* and *placeless spaces*, where instantaneous, technologised interaction is the norm. One outcome is the forging of new global identities and citizenships; yet virtualism also allows global corporations to spread their influence worldwide, and erode difference in the name of corporate profit.

It is thus remarkable that leftist commentators (such as Berman, Castells and Lefebvre) often depict technology as a 'cold' destroyer of authentic places. The implication here is that the city is constantly re-developed to 'keep up' with the ever-accelerating pace of life, and to accommodate new communication technologies. Each time this occurs, some things are inevitably lost for ever. Despite this, and the fact that some urban dwellers may 'resist' the onset of new technologies, most become more-or-less willing participants in technological innovation (witness the gradual normalisation of technologies such as home computers and mobile phones: once positioned as desirable luxuries, they are now overwhelmingly regarded as essential goods). The story is one of social regress accompanying technological progress: the quantitative gains enabled by new technologies result in qualitative losses and, possibly, the impoverished quality of the urban realm. In the informational age, for example, the rise of mediated communication is blamed for the decreased quality of public discourse, with the arts of face-to-face encounter lost to a generation weaned on email and text. Physical marketplaces, too, are seemingly declining in importance as e-retailing and eBay take over. Perhaps it is only a matter of time before virtual universities become the norm rather than the exception.

Of course, much too much is made of the eroding powers of new technologies, and it is conversely possible to argue that virtualism actually enriches the quality of physical encounter and interaction (W.J. Mitchell 2004, for example, makes much of the use of mobile phones as an enabler of 'approximeetings', whereby shifting social arrangements can be refined via text and chat). Furthermore, it is too easy to become fixated on the potential impacts of 'new' technologies on urban life, and ignore the multifarious ways city life relies on technologies both old and new. Graham and Marvin (1995) accordingly insist the contemporary pre-occupation with the informational superhighway should not detract from the other (largely neglected and taken-for-granted) networks and infra-structures that remain essential to post-industrial urbanisation. Likewise, Latour and Hermant (2001) point out the fragments of the city are meshed together by 'networks of speed, light and power' which are often untheorised and understudied (and hence which constitute an 'invisible city'). Such arguments emphasise the need for systematic analysis of the connections between contemporary urban life and networked urban infrastructures, including energy, transport and water networks. Sewage networks, for example, may be regarded as 'dull', 'banal' or 'boring', yet are one crucial network allowing social and economic power to be extended through urban space. Emphasising this, Gandy (2004b: 180) demonstrates that the flow of water through urban space remains as important today as in the earliest cities examined by Wittfogel (1957), and plays a pivotal role in freeing populations from disease, squalor and misery. Noting major shifts in the hydrological dynamics of the city over time, Gandy concludes the social organisation of water infrastructure demands careful attention in an era where it is becoming (at least in the West) subject to consumerist impulses and market deregulation. Clearly, focusing solely on electronic and digital infrastructures when water, gas and electrical networks remain vital to industrial and domestic life deflects attention from the myriad ways that 'old' and 'new' technologies entwine in the social reproduction of urban life.

ANIMATING URBAN TECHNOLOGY

Without doubt, the relationship between cities and technology is a fasci-nating one. However, in many accounts, this relationship is depicted as one-sided: technology is seen to drive urbanisation, with successive innovations triggering major changes in the form and function of cities. Very often these accounts lapse into *technological determinism*. One current candidate for this is the World Wide Web, which some commentators (including Castells) suggest is producing a new social morphology (and a new urban order). The reality is more complex though, as technology can never transcend the spatial and temporal constraints of the spaces and times in which it is embedded. Stephen Graham (2004a) flags this up when he discusses the complex ways in which new media technologies are being used in real ways, in real places. Citing the sociologists Haythornthwaite and Wellman's (2002) view that urban life and new media tend to be constituted together, he argues for research which moves beyond generalised and deterministic discourses about the 'impacts' of technology on society to look in rich empirical detail at the way these technologies are woven into the fabric of city life.

Rather than buying into ideas that technology shapes society, there is thus a significant turn in the social sciences towards exploring the entwining of the social and the technical (Thrift 1995). This recognises that technologies are not drivers of urban change in and of themselves, and co-evolve with the urban fabric as they become woven into the social, economic and political life of cities (Graham and Marvin 2001). At the same time, this approach recognises the work that is required to incorporate new technologies in the life of the city. By way of example, Graham and Marvin (2001) underline that the World Wide Web is not endowed with causal powers (i.e. its invention did not cause cities to take new virtual forms). Rather, the Internet has transformed urban life inasmuch as it has been incorporated into a multiplicity of complex networks of modems, servers, software, telephone connections and computers, all of which are of course dependent on reliable electricity infrastructures. Further, it is clear that all these networks need to be *engineered* in a variety of ways, and that for all the talk about digital revolution, much effort needs to be expended threading IT networks under the busy roads and pavements of the urban fabric, to the smart buildings, dealer floors, headquarters, media complexes and stock exchanges that are the most lucrative target users. Consequently, Graham (2004a) estimates fully 80 per cent of the costs of a network are associated with this traditional, messy business of getting it into the ground in highly congested urban areas.

This perspective on the networked nature of urban technology is elaborated at length by Graham and Marvin (2001) in *Splintered Urbanism*. This volume describes the way in which modern capitalist societies have come to rely on a whole interconnected web of infrastructure networks of electricity, water, gas and sewage, as well as automobile transport systems, rail and air transport, telecommunications, the Internet and media networks. These networks, they argue, tend to accrue in society on an incremental basis, creating ever denser and more expansive city infrastructures. Consequentially, they suggest that a holistic, trans-disciplinary and robust understanding of the city as a critical amalgamation of infrastructure networks is required if we are to understand the relations between cities and technologies (and the ever-attendant threat that this relationship will break down if networks fail). Indeed, the fact that the spatial formations of city life (including infrastructures of utilities, transport and communication) are increasingly controlled by software means that if the code crashes, tasks cannot be completed (Dodge and Kitchin 2005). Moreover, given that software is both replete with international conventions (such as communications protocols) yet mutually anarchic (think of the debates of Mac versus Windows operating systems), problems of translation and compatibility need to be constantly worked through (MacKenzie 2005). As such, the relationship between human and technology is contingent, relational and complex. Operationality depends upon the constitution of collective agency,

In theoretical terms, the acknowledgement that there are complex and constantly unfolding connections between technology and society owes much to the work of those who are active in the study of science, technology and society (STS). Broadly, the aim of STS has been to offer an alternative perspective on the practices of science and the ways in which scientists represent their findings. Its most celebrated proponent, Bruno Latour, produced an understanding of the process of scientific discovery (*Science in Action* 1987) which portrayed scientists as actively engaged in the pursuit of political and economic influence, as well as academic acclaim. One of the ways he suggested they do this is by creating simplified areas of knowledge which they present as self-evident: Latour terms these 'black boxes'. These black boxes are assumed to be true, but left unexamined, allowing scientists to continue to construct theories and ideas on the basis of knowledges which are relied upon but rarely subject to critical analysis. Latour argued scientific 'progress' is based on the addition of many black boxes (see Laurier 2004). However, Latour asserted that this construction of knowledge does not guarantee acclaim, arguing that scientists need to undertake a range of actions to convince people of the worth of their ideas (see Latour 1987). This involves them 'crafting' experiments, using humans and non-humans as mediators who then vouch for the veracity of these facts both within and beyond the laboratory setting.

Latour hence used his ethnographically derived insights about scientific activity to make a more general point about technology – namely, it relies upon unexpected connections between heterogeneous elements. Significantly, by adopting an 'open' approach to the study of social phenomena (i.e. viewing the world as a 'relational network' of objects, people and representations), Latour extended notions of agency to non-humans, with the performative production of power through network interactions seen as involving the enrolment of *actants*. Accordingly, Latour's work is frequently cited as influential by those working with *Actor Network Theory* (a term he actually describes as 'rather silly' – Latour 2003: 35) (Box 4.2). In essence, Actor Network Theory (or ANT) offers an analytical framework for making sense of social relations and struggles of all kinds, avoiding over-generalisation or the search for 'deep' structures in favour of descriptions of the associations that bring society into being (see, for example, Law 1992; Latour 1987, 1993). Latour stresses that these networks do not actually exist as such, but are the paths traced by a researcher trying to make sense of any scientific knowledge, idea or concept. An example he offers is BSE infection (Latour 2003): while one might begin by considering the way that scientists in laboratories test meat for the presence of the prions blamed for the outbreak, tracing the origins of the concept of BSE would involve reading government reports, visiting abattoirs, hospitals and scientific establishments and talking to bureaucratic elites.

Accordingly, ANT suggests that if we want to understand the impacts of new technologies on urban life, we need to explore how the social relations that produce these technologies are enabled and sustained. Law (1999) argues that this is very different to suggesting technologies determine social life: rather, it is to assert that technologies are worked upon and incorporated into cities in a variety of ways, creating fragile interdependencies. Further, opening up questions of technology to consider the connections made between social actors and non-social elements begins to shed light on the way in which different materials – objects, bodies and texts – are brought together to produce particular outcomes.

Crucially, ANT implies power does not belong to particular groups, politicians or individuals, with social effects produced by a combination of heterogeneous entities (even though it might be attributed to one of

Box 4.2 ACTOR NETWORK THEORY

Actor Network Theory originates within studies of science, French intellectual culture and, above all, the writing of Michael Serres and Bruno Latour (see Bingham and Thrift 2000). In many ways, its focus on networks transcends the dichotomy of micro- and macro-level analysis that characterises social research, insisting on an approach which neither distinguishes structure and agency, nor differentiates local and global. Further, it provides an account of social life which incorporates non-human elements as *actants*, on the grounds that they can be just as important as human actants in making things happen. For example, universities consist of more than social relations between students, lecturers and ancillary workers: these social relations are held together by all manner of information networks, databases, proformas, departmental offices, websites, lecturing rooms, overheads, texts, cafeterias and so forth. It is the gathering together of these diverse social materials in topological networks that ANT focuses upon. In short, ANT is based on a distinctive ontology or belief about how the world works – namely, that it involves the co-relation of all manner of things, and not just people. The idea that seemingly incompatible materials need to be treated in a similar manner distinguishes ANT from classic Marxist and structural readings of social life as well as humanist interpretations. While some geographers have insisted that ANT provides a suitably spatialised perspective on the making of society through networks (e.g. Murdoch 1998), others have been more critical (e.g. Peet 2005: 166 describes ANT as 'the biggest fraud ever visited on social theory').

Further reading: Law and Hassard (1999)

them). As Murdoch and Marsden (1995) argue, the powerful are not necessarily those who *hold* political power, but those who enrol, enlist and coerce people and things into associations that allow policies to be enacted. These networks are not spatially constrained, so that a network that allows policies to be pursued in a specific city may involve representatives of global corporations, national media companies and state politicians as well as local actors (see Chapter 5).This means that actors can often exercise influence at a distance, and that those located in the centre of a network may be physically distant from those with whom they are associated. Networks can be 'fat' or 'thin', 'deep' or 'shallow'; their effectiveness does not depend on their size or shape but the extent to which they create the ability to act.

Ultimately, ability to act relies upon the type and strength of association between different actors, something that may rely on the representations that give the network a sense of identity or purpose (see Chapter 2 on city representation).This emphasises that materials, technologies and texts are a crucial part of networks, and that 'actants' in networks may be non-human as well as human. For example, 'town plans' are just as enmeshed in planning networks as urban planners.When enrolled in networks, these actants hold each other in positions that define the identities of subject/ objects and their ability to do things (Murdoch 1997). Accordingly, examination of the circulation of things and people around networks is an essential feature of ANT. In the context of urban technologies, this means considering the associations between information, discourses, machines, documents, data, ideas and representations as much as it involves thinking about the social relations between technocrats.

ANT has ushered in a new language of 'actants' enrolled in the networks of 'immutable mobiles' which bring the world into being. Open to notions of complexity, and mindful of the fact that the world is always in flux or in the process of becoming, ANT thus avoids making totalising claims (meaning that it is very much in keeping with post-structural thought). As Whatmore (1999: 27) argues, for those working with ANT, 'agency, and by implication, power, is decentred, spun out between social actors rather than seen as a manifestation of unitary intent.' Moreover, given that Michael Serres and Bruno Latour have used geographical metaphors frequently in their work, ANT seems to offer a distinctively spatial way of describing, rather than understanding, the world. Latour and Hermant (2001) give a flavour when they begin to consider the role of Paris's street furniture in the life of the city:

> Paris has on its ground roughly 770 pillar-shaped billboards, 400 newspaper kiosks, two theatre kiosks, 700 notice boards 2000

advertising billboards, 400 street urinals, 1,800 bus shelters, 9,000 ticket machines, 10,000 traffic lights, 2,300 letter boxes, 2,500 phone boxes, 20,000 baskets, 9,000 benches . . . Each one of these humble objects, whether a urinal or basket, tree surround or street name, phone box or street light, forwards, by its color or form, its practice or position, a particular injunction, authorization or a prohibition, a promise or a permission.

(Latour and Hermant 2001: 12)

If, as Latour and Hermant (2001) contend, Paris is conceived as a network, then its *power* is diffused through an array of seemingly mundane objects, each of which is, nonetheless, part of the whole. Equally, each plays a part in a complex and constantly mutating city-assemblage that holds together through the affordances and interactions of objects. They illustrate this by talking of the role of the iron tree-surrounds that both protect Paris's trees and provide the ideal place for chaining up bicycles, the seatless bus shelters which shelter commuters from the rain but offer no refuge to the homeless, or the pieces of rolled carpet that guide water around the gutters to clean the streets (Plate 4.2). Each is part of Paris, and although it is customary to start any exploration of power in the city at the town hall or council chamber, Latour and Hermant (2001) insist that we need to take the agency of objects seriously too. After all, who is to say what really makes a city work in the way it does?

From an ANT perspective, it is thus possible to see that new technologies do not drive urban change, but are caught up in complex networks (or socio-technical *assemblages*) which incorporate all manner of actants. We can explore this contention further through an exploration of the rise of surveillance in cities – a topic widely discussed in contemporary urban geography. As Koskela (2000) notes, electronic surveillance has become commonplace in both public and private spaces. For instance, surveillance cameras are commonly used to protect high-class private premises – 'gated communities' (or privatopias) – as well as semi-public places such as shopping malls, train stations, university campuses, leisure centres, libraries and airports (Adey 2004). What is of particular interest to many commentators is that the use of closed circuit TV (CCTV) surveillance now extends to spaces traditionally thought of as entirely public, such as streets and parks (Fyfe and Bannister 1996). Almost every town centre in the United Kingdom now has extensive CCTV surveillance

Plate 4.2 An urban actor? Carpet in the Parisian gutter (photo: author)

systems, and such systems are also proliferating in many suburban and residential spaces (Plate 4.3).

When coupled with speed cameras, infra-red car number plate recognition, mobile phone tracking, digitally mediated consumption

Plate 4.3 CCTV has become a routine (and more or less accepted) part of urban life (photo: author)

systems and credit checking mechanisms, CCTV systems effectively transform the landscape of the contemporary city into a militarised *scanscape* (Coaffee and Wood 2006). Justified as a means of crime reduction, many commentators suggest these new technologies are not just about crime control, but control *per se* (Fyfe and Bannister 1996). One of the main concerns voiced by those anxious about the rise of CCTV is that, post 9/11, many cameras incorporate Neural Network Face Recognition software which may be programmed to track people with specific physiognomic features (Lyon 2003). The potential for such systems to be used in discriminatory ways has been widely noted; for example, it has been claimed that many of these systems are programmed to track non-white individuals. Importantly, digital systems also allow for algorithmic surveillance, whereby surveillance systems use software to move from classification and comparison to prediction or even reaction (e.g. a CCTV system may trigger an alarm on seeing a specific individual entering a particular area). Countering allegations that CCTV may be used for discriminatory ends, the developers of such algorithmic systems claim these actually reduce the potential for discrimination. For example, a racist police officer cannot decide to arrest any non-white male when a facial recognition system can decide categorically whether a particular individual is the wanted individual.

Lyon (2002) argues that as electronic surveillance becomes intelligent and immanent within the city, notions of traditional disciplinary control are being replaced by the continuous electronic disciplining of subjects against predefined norms (see also S. Graham 2005). For some, this raises fears about the extent to which authority is handed over to 'machines'. Lyon (2002) cites an extreme example, namely the development of movement recognition software linked to an automatic 'lethal response', potentially killing intruders without explanation or appeal. A more routine example is a shopper being unable to purchase goods on a store card because their credit rating does not fit the store's profile: arguing with the shop assistant that the machine is wrong simply does not work, because they have no authority to override the system. Graham (2005) also talks of the way that telephone queuing systems might filter calls, prioritising those from specific areas or allowing those with positive credit ratings to queue jump. Such examples raise the spectre of cities being controlled by a faceless and unverifiable technology, with technology supplanting human authority. For such reasons, many individuals have a pathological distrust of electronic surveillance, forgetting that surveillance is ultimately about the imposition of human norms and behaviours. Koskela (2000) underlines this when she talks of the agency of CCTV cameras:

> [E]ven if the camera seems to look down from above, *the camera itself has no eyes*. Its lens is blind unless someone is looking through it. The

camera seems to be looking at people from above but the monitoring room may be, for example, in the basement of a shopping mall where premises are cheaper . . . This makes it very difficult to ask for help through the agency of the camera – the camera *leaves its object entirely as an object*: passive, without any ability to influence the situation.

(Koskela 2000: 249, original emphases)

CCTV technology may not be discriminatory in itself, but its operators and programmers may be (as Norris and Armstrong 1999 confirm). The point here is that, like other technologies, CCTV systems need to be animated through their incorporation in a surveillant assemblage which potentially includes police officers, camera operators, engineers, programmers, retailers, security guards and politicians (as well as a diversity of hardware and software). Although it is possible to argue that these assemblages are less and less mediated by human discretion, it is quite wrong to suggest technology is becoming all-powerful. Rather, it is a case of constantly questioning how technology and society align to create more or less stable networks of regulation, surveillance and control.

NATURE AND POST-HUMAN CITIES

In historical terms, human civilisations are deemed to have developed and evolved by distancing themselves from nature and all that is wild. Hinchliffe (1999) refers to this as a *foundational story* of cities: a story wherein nature is gradually expunged from cities as the city grows and develops over time. Nature is thus held up to be the antipode of civility and culture:

The wild occupies a special place in the imagined empires of human civilisation, as that which lies outside its historical and geographical reach . . . [it is] a place without us, populated by creatures (including surreptitiously uncivilised humans) at once monstrous and wonderful, whose strangeness gives shape to whatever we are claimed to be.

(Whatmore and Thorne 1998: 435)

Writing of the processes of boundary-making that separate nature and culture, and thus define the essence of civility, Kay Anderson (2000: 312) demonstrates that the struggle for Selfhood requires a fundamental disengagement and distancing from the 'animal'. There is thus a rich history of manners that dictates that civilised human beings do not act as animals do: it is hence impolite to belch, fart, defecate or spit in public; public nudity and sexual display are frowned upon; overindulgence and binge drinking are seen as giving in to animal urges (Schilling 2005). The corollary of this is that we have evolved elaborate social rituals of courtship, cleansing, eating, drinking and dressing designed to indicate our conquest and repression of nature.

Although often depicted as an inner struggle (i.e. a battle to restrain our animal urges), Kay Anderson also alerts us to the way culture–nature distinctions take material form in the topographies of the city. For example, her examination of the Royal Zoological Society of South Australia suggests that urban zoos serve as domesticated spaces within which 'irrational' nature can be contained and displayed. In her estimation, within zoos and menageries, 'the raw material of nature is crafted into an iconic representation of human capacity for order and control' (K. Anderson 1995: 278). Confined within paddocks and cages separating non-human animals from human animals – and non-human animals from one another – 'wild' animals have increasingly found their occupation of urban space limited to specific sites of visual spectacle. Anderson suggests this process of domestication was part and parcel of processes of modernisation: as cities came to be read as monuments to people's capacity for progress and order, then zoos represented the ultimate triumph of modern man (sic) over nature, of city over country, of reason over nature's apparent wildness and chaos. Wolch (2002) likewise argues that human relationships with other species are absent in modern ideals based on a notion of progress rooted in the conquest of nature by culture.

The significance of fears of animality and incivility has also been manifest in other policies of urban improvement and modernisation. In the rapidly industrialising cities of the nineteenth century, for instance, the mantra of civic improvement was accompanied by concerted efforts to purify the city, with the cleansing, paving and lighting of streets viewed as a precursor to the creation of commodious and civilised city life (Ogborn 1998). The exclusion of animals from the modernising city was a crucial part of this process, with Philo (1995) suggesting that newly forged moral and medical knowledge triggered attempts to remove cattle slaughter, milk production and other 'noxious' trades from the city. In the process, it was hoped to eradicate the odours, flies and unseemly sights associated with livestock from intruding into the streets. At this time then, while certain animals were seen as acceptable on the streets as pets (e.g. cats, dogs) or working animals (e.g. horses), pigs, sheep and cows were deemed as 'out of place' and banished to the countryside. The implication here was that having animals marched through the streets was incompatible with health, safety and commerce, with some reformers even making a connection between the sight of cows rutting in the street and the 'improper' sexual conduct of the working classes. Debates about slaughterhouses crystallised these issues, given that butchery was seen as introducing putrid, rotting product onto the street, with blood, offal and dung circulating in the gutters. In London, for instance, new by-laws restricted the proximity of slaughterhouses to public streets and thorough-fares, with the construction of the new Smithfield market in 1868 providing a covered facility with direct railway access. Similarly, in Paris, the slaughterhouses were moved out to La Villette, from where carcases could be transported into the central city by both canal and road.

Yet it is not only animals of various kinds who have found their occupation of the city under threat. Other forms of nature – trees, grass, plants – have been increasingly controlled and domesticated, allowed to grow only in designated spaces such as urban parks or domestic gardens. The fact that these spaces are manicured and highly managed underlines the precarious position of nature with cities: unwanted flora ('weeds') are relentlessly extinguished, while plants are pruned, trimmed and cut so that they do not 'outgrow' their setting. In contemporary cities, there-fore, the culture–nature distinction remains crucial in configuring the boundary between public and private space. Kaika (2004) thus identifies the home as a space designed to keep natural processes and elements (e.g. dust, cold, polluted air, dirt, rain, sewage, smog) outside, albeit this relies on a complex distinction of 'good' and 'bad' nature. Clean, processed water is then regarded quite differently from rain or river water, and thus only certain types of water are granted admission to the household through plumbed networks.

The *selective porosity* of the home is an indicator of the way that we seek to maintain a distinction between nature and culture: nature is allowed to intrude on the domestic only if it is presented in a civilised, manageable and commodifiable form (indeed, nature persists in the domestic not just in the form of flows of processed water and conditioned air, but also in the shape of pets, house plants and cut flowers). Nevertheless, Kaika

(2004) stresses that all of these – like the purified water that runs through the home – are in fact hybrid products of an interaction between the physical environment and human activities. Swyngedouw (1999: 445) echoes this when he speaks of cities as networks of interwoven processes that are simultaneously 'human, natural, material, cultural, mechanical and organic'. As such, the myriad of processes supporting and maintaining domestic life (for example, the supply of water, energy, food and shelter) always combine society and nature in particular ways.

This attempt to maintain the boundaries of the home as a divide between inside and outside may seem quite unremarkable, but several geographers have suggested that the persistence of the nature–culture dichotomy as a basis for constructing notions of Self bolsters iniquitous geographies on different scales (see especially Sibley 1995). By way of example, in her wide-ranging study of post-war French society, Kristen Ross (1994) makes an explicit link between bodily cleanliness, home improvement and the French desire to repress colonial insurgency. The title of her book – Fast Cars, Clean Bodies – hints at the ways in which French anxieties about the Algerian war coincided with dramatic rises in spending on vacuum cleaners, fridges and consumer durables, all of which promised cleanliness, order and efficiency. Here, rituals of domestic and bodily purification designed to eliminate polluting matter paralleled French geopolitical strategies of containment and repression in North Africa. Developing similar arguments, Kay Anderson (2000) argues that the social construction of certain racialised groups as uncivilised can be seen as one way white fears of the animal are assuaged, with anxieties about the integrity of Self projected onto racialised Others, who are then repressed and contained. However, this ascription of wildness to certain bodies is by no means unchanging (as Anderson shows through her examination of 'aboriginal' housing projects). Bodies are not self-evidently different or racialised: rather, they become marked as coloured through socio-spatial inscriptions (with, for example, certain white groups labelled as not-quite-white because of their alleged closeness or affinity to nature – see Sibley 1995 and S. Holloway 2005 on the exclusion of gypsy-travellers in the urban West).

But what is perhaps crucial about the dichotomy of city–nature is that it is never so clear cut as we would perhaps wish it to be. Though repressed, nature always returns. One only need look at the way that abandoned buildings soon become host to a number of birds, plants and insects to appreciate that nature can flourish in cities (Plate 4.4) (Edensor 2005). In any case, there are many animals that live, with varying degrees of success, in cities or urban borderlands. Wolch (2000) suggests that in North America alone, these include bears, cougars, coyotes, raccoons, skunks, squirrels, foxes, deer, starlings, reptiles, amphibians and colonies of feral cats. In ecological terms, this means that cities often provide more biodiversity than rural areas (especially when these are intensively farmed). Moreover, despite efforts to eliminate them, species including rats, feral cats and pigeons have become more widespread in urban than rural spaces, adapting to the built environment and feeding on the detritus of urban life (recent UK media stories even report crack-squirrels, addicted to the residues of drug-taking). Hinchliffe (1999) refers to these as *city-nature formations*, adaptive species that could not survive in the 'wild'.

Griffiths et al. (2000: 58) thus conclude that despite modern urbanism's preoccupation 'with the *elimination* of unregulated nature . . . the realisation of an ordered city is an impossible project'. The impossibility of making a neat distinction between culture and nature has also been emphasised by the emergence of cyborg bodies, artificial intelligence and robotics. Long the subject of science fiction, the idea that

Plate 4.4 The return of the repressed? Nature in the city, reclaiming an abandoned industrial site (photo: Tim Edensor)

the biological body can be supplemented or conjoined with technological apparatus in ways which extend its capabilities is becoming a reality. For instance, recent advances in biotechnology, genetic engineering, cloning and prosthetics point to the ways that technoscience increasingly works with biological materials in ways which erode conventionally held distinctions between the human and the non-human, between the biological and the technical and, again, between nature and culture (L. Holloway 2004). Moreover, new wearable communication devices allow people to become anchored into the digital infrastructure of the city – to the extent that billboards can register the presence of an individual and communicate directly with them (W.J. Mitchell 2004). The figure of the *cyborg* is thus central to the work of Donna Haraway, who has offered a number of insightful commentaries on the possibilities and paradoxes of hybridity, not least in the realm where science, nature and the city collide (see Box 4.3).

Box 4.3 CYBORG CITIES

For many years, the cyborg – exemplified by the Cybermen in TV's *Doctor Who* – was the stuff of science fiction. Essentially a fusion of the human and the machine, cyborg bodies have nonetheless become commonplace because of developments in biotechnology and communication science which have allowed for computer implants, electronic prosthetics and wearable communication devices (Schilling 2005). The potentiality of such bodies to refigure the relationships between nature and science has been widely noted, as have the unpredictable consequences of replacing the biological with the mechanical. Particularly significant here is the work of Donna Haraway, who has noted that both communications sciences and modern biologies are constructed by a common desire – the translation of the world into a problem of *coding*: 'a search for a common language in which all resistance to instrumental control disappears and all hetero-geneity can be submitted to disassembly, reassembly, investment, and exchange' (Haraway 1991: 57). She thus alleges that high-tech culture challenges existing dualisms such as self/other, mind/

body, culture/nature, male/female, civilised/primitive, reality/
appearance, effectively removing distinctions in the search for
equivalence and translation. Her conclusion – that cyborg imagery
can suggest a way out of the maze of dualisms by which humans
have understood their bodies and machines – thus supports a
feminist politics in which repressive gender identities are dissolved. While others have ignored the radical gender potentials of
cyborg bodies, there is a widespread recognition that the forms
of life which the architects and urban designers of the future will
plan for will be very different. Elaborating, Gandy (2005) concludes
that if we understand the cyborg to be a cybernetic creation, a
hybrid of machine and organism, then we need to conceptualise
urban infrastructures as a series of interconnecting life-support
systems. At present, for example, Gandy (2005) describes the
modern home as a complex 'exoskeleton' for the cyborg body,
providing water, warmth and light; in time, the electronic extension
of the body's capabilities through e-communications and mobile
telephony will surely produce new architectures, with the entwining
of nature and technology producing cyborg cities where current
notions of the body and technology (or public and private, for that
matter) will no longer hold (see W.J. Mitchell 2004).

Further reading: Schilling (2005)

Because of such technological breakthroughs and innovations, the
contemporary era is seen as one where established boundaries of science
and nature no longer hold, with media panics over organ harvesting,
genetic profiling, animal implants and cloning indicating the profound
anxieties associated with new entwining of nature and culture. Yet it is
not only the blurring of 'natural' corporality and technology that raises
the spectre of hybridity. Indeed, advances in computing and artificial
intelligence mean that machines are also taking on a life of their own.
Writing of this, Thrift (2003a) contends that a number of different factors
are conspiring to produce environments that are full of 'machinic intelligence'. One is that the geography of computing is changing shape given

the lack of restrictions on where computing devices can be located (thanks to wireless computing). Thrift also suggests that the time of computing is changing, with computers able to run continuously and able to constantly interface with users. This is manifest in the advent of 'ubiquitous', 'pervasive' or 'everywhere' computing – constantly there, in the background, and no longer requiring our full attention. Related to this, Thrift reasons that computers are becoming more location aware, knowing where they are in relation to users and other devices, and are hence able to interact. This implies that computing is becoming more and more connective, with the purpose of computing devices being to communicate not only with the user but also with other devices. Connected to this, Thrift argues that the advent of 'soft' computing – i.e. based on algorithms that have a capacity to 'learn' – will mean that computers increasingly 'second-guess' the user, becoming a part of 'how they decide to decide' (Thrift 2003a: 392).

The ubiquity of machinic intelligence arguably presages new forms of everyday life. However, Thrift (2003a) illustrates his arguments with reference to the robots sold as toys yet which display sophisticated capacity for learning and interacting with their surroundings. While toys may not seem to be particularly important in the overall scheme of things, Thrift draws attention to them to underline the increasing difficulty of attributing agency. Callon et al. (2001) concur, arguing that all manner of suppressed 'things' – animals, bodies, codes, devices, information, documents, proteins – have forced their way into the company of the social in ways that disturb the accepted boundaries of human and non-human (as well as disciplinary boundaries in the arts and sciences) (see also Wolch et al. 2003 on the 'reanimation' of cultural geography). The implication here is that we need symmetrical theories of the city, in which no special emphasis is placed on either the human or non-human: all have the potential to make things happen. As Murdoch (2003) contends, space must be regarded as co-constructed by social and natural entities.

The assertion that nature and culture do not merely coexist, but inter-mingle to create new forms of life, problematises many of the assumptions that underpin urban theory, which has tended to work from the assumption that nature precedes (and is subsequently destroyed by) the city. As such, in the same way that notions of civility have allowed little room for nature, urban theory has tended to expunge nature. Rather than offering theories which speak of the fecund relationality of natures and cultures, urban theory has tended to operate with a more conservative definition of agency: namely, that people make cities (although, according to Marx, rarely in circumstances of their own choosing). Challenging this, Actor Network Theory (ANT) has served as an important reference point in the emergence of 'post-human' geographies. As discussed above, ANT supersedes all versions of a culture–nature binary, that is, divides between the social and the natural or the social and the technical, in favour of a conception of the world as inhabited by hybrids or quasi-objects that are not quite natural and not quite social. This rejects theories that privilege humans as significant actors. Agency is instead deemed a relational effect generated in networks by humans and non-humans alike.

In general terms, ANT is a theory that seems tailor-made for a world made up of hybrid objects, heterogeneous networks, and fluid identities, allowing for the identification of new ('distributed') forms of subjectivity (Murdoch 2004). Adapting the principles (if not always the language) of ANT, geographers are beginning to grapple with the multiple forms of life that animate cities: not only the technologies, screens, software, calculations and knowledges that constitute the virtual city, the engineered city of buildings, roads, tunnels, sewers and cables, or the repressed city of weeds, parasites, animals, ghosts and monsters, but also the hybrid forms that transcend these categories.

One interesting line of inquiry here is that which explores the making of 'city-nature' hybrids, such as the smallholding (L. Holloway 2002), the allotment (Crouch 2003) and even the local cemetery (Cloke and Jones 2004). In each, urban green space becomes a practised 'formation':

> a setting where the hobby farmer, the plotter, the vegetable grower, the artist, the dog-walker, the dog, human rambler and the fruit harvester are encountered in passionate, intimate and material relationships with the soil, and the grass, plants and trees that take root there.
>
> (Lorimer 2004: 90)

In such accounts, the agency of non-humans is as important as that of humans: agency is 'spun out' between different actants. Cloke and Jones (2004) illustrate this with reference to the role that trees have played in creating a distinctive place identity for a cemetery in Bristol (UK). Noting that several of the ornamental trees propagated in Victorian times have self-seeded in the cemetery, they suggest:

> the nature of the cemetery is not an inert surface inscribed by social forces, but an unruly force which has pushed back against human intentions, colonising not only the spaces but also the times when intentions are weak, ambiguous or in conflict.
>
> (Cloke and Jones 2004: 327)

Although Cloke and Jones (2004) admit to some difficulties in imbuing trees with agency, their conclusion that involvement of nature in the relational agency and dwelling of place points to the need to trace the effects of non-human agents. Furthermore, Whatmore and Hinchliffe (2003) suggest that this way of thinking about city natures might also provoke new ways of fostering urban ecologies. Far from being a contradiction in terms, they suggest that urban biodiversity is starting to be accorded the kind of conservation significance once reserved for rural and sparsely populated regions. As such, significant energies are beginning to be invested in what they term 'recombinant ecologies' – 'the biological communities assembled through the dense comings and goings of urban life, rather than the discrete and undisturbed relations between particular species and habitats that are the staple of conservation biology' (Whatmore and Hinchliffe 2003; Hinchliffe et al. 2005). As they document, there is growing awareness of the importance of this 'recombinant ecology' as a component of what makes cities liveable. One manifestation of this is the rise of urban nature projects (such as Groundwork UK's Living Spaces programme) not merely concerned with beautifying public spaces through planting, but protecting abandoned railway corridors, derelict buildings and canals as prodigious city-nature sites. Another is the European 'Slow Cities' movement, which articulates a vision of civilised urbanity based on a slower pace of life, tranquillity and the reintegration of nature in a variety of forms.

In the final analysis, such attempts to problematise taken-for-granted notions of human agency offer new takes on the making and unmaking of urban landscapes. They also inject a note of caution into any discussion of nature, since it is harder and harder to find any form of 'pristine' nature that has not long been incorporated in social processes of production and consumption. Yet the dissolution of the boundaries between nature and society may come with a price, as Davies (2003) spells out:

Our explorations of hybrid geographies may have the potential to both contest and facilitate the aesthetic and genetic mutability of animal bodies and future natures. In tracing these transgressive nature-cultures we thus need to think whether and where we might want to redraw boundaries, to identify the ethical contexts or aesthetic moments in which we might want to keep agencies, entities and animals from becoming too entangled, without simply reasserting a purified human-animal divide. In the challenge to envision a geography that is beyond the human–nature divide we need to be sensitive to these complexities, aware of our responsibilities to the multiple agents constituting these terrains, and the politics of our practices in representing them.

(Davies 2003: 412)

As Davies implies, we may wish to dislocate established ideas about the centrality of human subjectivity, yet geographies will inevitably continue to be written by people who will remain, by and large, certain of their place in the world. Much as we may like to develop 'post-human' geographies, we therefore will continue to impose our ideas of agency and subjectivity onto actants who cannot write their own geographies. As such, while theories of hybridity suggest the heterogeneous networks of city life include all sorts of 'wild' things, whether we can ultimately develop symmetrical accounts in which machine, animal, plant and human lives are registered in similar ways remains open to debate.

CONCLUSION

In an era where the software and hardware of the city are producing new spatial formations, cities are seen to have become extraordinarily intricate (Amin and Thrift 2002). This perceived complexity is encouraging geographers to consider the relationship between society and technology in new ways – albeit that many retrospective analyses suggest that this profligate combination of social and technological materialities has always been a defining feature of the urban condition. Likewise, drawing on new ideas about culture, nature and subjectivity (many of which have been developed in rural rather than urban studies), geographers have started to explore the entwining of nature and culture in the making of space, posing new questions about the nature of individual as well as collective identity, the human–animal divide, and the nature of subjectivity and agency itself.

Emphasising the recombinant ecology of the city, and simultaneously problematising the distinction of nature and culture, such new takes on the city unsettle the anthropocentric certainties of theory and throw new variables into the urban mix. Clearly, grasping the effectivities and agencies associated with technologies, materials and animals as they enter into relationships with humans requires the use of methodologies which are attentive to the multiple forms of life which exist in cities. Yet the task is much wider than this, with attempts to 'reanimate' the city requiring the adoption of theoretical frameworks (such as ANT) which challenge our ontological understanding of how cities work. The literatures considered in this chapter therefore map out some important directions for the development of post-human geographies (or at least, the creation of geographies where non-human agency is given serious consideration). However, it is apparent that much work remains to be done if urban geographers are really to reanimate the city in both theory and practice, and to develop accounts doing justice to the hybrid and multiple materialities of cities.

FURTHER READING

There are some fascinating and textured accounts of the introduction of particular urban technologies: Wolfgang Schivelbusch's (1986, 1988) work is much acclaimed, particular his consideration of railway journeys (which changed senses of space and time) and his history of the industrialisation of light. More extensive are the three volumes of the Cities and Technology series (Chant and Goodman 1998, 2000; Roberts and Steadman 1999) (Open University and Routledge) which focus on pre-industrial, European and North American cities respectively. Graham and Marvin's (2001) Splintering Urbanism is an obviously important reference on infrastructure networks, while Graham's (2004c) Cybercities Reader is a useful primer on Internet geographies. William Mitchell (1995, 2004) has produced some of the more utopian accounts of the possibilities of cyborg urbanism, but his work certainly provides food for thought. Those wishing to explore debates surrounding the production of nature are advised to refer to Noel Castree's (2005) volume on *Nature* in the *Key Ideas in Geography* series; among recent publications by geographers, Sarah Whatmore's (2003) *Hybrid Geographies* represents probably the most important intervention in these debates.

5

THE INTRANSITIVE CITY

A path is always between two points, but the in-between has taken on all the
consistency and enjoys both an autonomy and a direction of its own.

(Deleuze and Guattari 1987: 380)

Questions of scale have always been fundamental to the study of human
geography, with many of the key concepts in the discipline intended
to divide the world up into manageable units for the purposes of analytical convenience. On this basis, geographers have invoked a variety of
territorial units which serve to guide research and pedagogy, such as the
neighbourhood, the city, the locale, the region, the nation state and so
on. What is particularly important about these concepts is that they are
regarded as being nested in one another like a Russian *matryoshka* doll: the
neighbourhood is in the city, the city is in the region, the region is in
the nation and so on. The idea that social and economic processes operate
in different but related ways at different scales has hence constituted a
keystone of geographical inquiry.

As such, a hierarchical and Cartesian notion of scale is fundamental
to the geographic imagination, and has allowed geographers to develop
distinctive ideas about how changes on one scale inform processes operating at others. For example, writing in the context of political geography,
Taylor (1985) has suggested that capitalism is legitimised in distinctive
ways at different levels with the world economy, nation state and locality

providing systemic integration in an ideological sense. Likewise, the much-vaunted *localities* debates of the 1980s explored the uneven impacts of global economic change on different cities and regions, asserting that while the global economy provides the context for urban and regional change, the uneven development of localities conversely drives processes of capitalist accumulation (see Cooke 1989). Here the world was conceived as a patchwork of places, the city region seen to be located within national and global economies.

Yet such taken-for-granted notions of scale have recently been problematised by a geographical literature that considers the production of scale. This effectively takes the focus away from either the national, local or global as the starting point for explanation in geography, suggesting that each must be regarded as socially produced (Swyngedouw 1997; Brenner 1998). This literature carries some profound implications for the examination of the city, encouraging many urban researchers to question where to locate cities within extant hierarchies of scale. A key emphasis here is on the *intransitivity* of cities – the idea the city exists on multiple scales simultaneously, and does not simply take its allotted place in a system of space that extends from 'large and important' (i.e. the global economy) to 'small and trivial' (i.e. the spaces of everyday life). In part, this emphasis on the pliability and fluidity of city space is associated with new understandings of globalisation which have impelled urban theorists to reconsider the place of cities in the world. Reviewing some of this work, this chapter alights on literatures which identify the city as caught up in complex global chains of production and consumption. Recognising the pivotal role of cities in articulating global networks, this chapter therefore considers how we might better locate the city in a mobile space of material flows. Exploring notions of connectivity and mobility, the chapter concludes that the literature on world cityness provides some important clues as to how we might rescale theory to better acknowledge the spatialities of the city. The chapter starts, however, by focusing on the idea that we live in a world where 'everything flows' and sedentary theories no longer apply.

KEEP ON MOVING: HYPERMOBILE CITIES

In previous chapters, some of the ways that movement and mobility are embedded in cities through (for instance) the provision of transport systems were discussed. By way of example, we explored the ways that car travel is normalised through an infrastructure of roads, car parks, petrol stations, motels and drive-ins, which exist in symbiosis with a society of motorists and passengers. But movement is evident in a multiplicity of urban formations and practices: the movement of data through computer infrastructure and communication networks; the circulation of news through radio transmissions, newspapers and TV broadcasts; personal communication via phone and text; the flow of sewage through drainage pipes, daily deliveries of mail and so on. Yet what is increasingly evident is that we live in a hypermobile era where these flows are both *speeding up* and *spreading out* (Adams 1999). In relation to the latter, one can simply refer to the rising number of business travellers who routinely commute to work by air travel to make the point that urban networks are becoming more spatially dispersed (with Sudjic 1991 suggesting that the airport has become perhaps the single most important entry point to the city). The year-on-year increase in passenger movements is testament to this (see Figure 5.1): the fact that more and more passenger movements are intercontinental rather than within one nation suggests that distance is becoming less of an obstacle to movement.

Quick, cheap and regular air travel is one of the major enablers of *space-time distanciation* – the process whereby more and more social relations are routinely maintained at a distance, so that, for example, intercontinental business travel and transaction are regarded as commonplace. Giddens (1990: 29) contends that this stretching of social relations has produced a 'global present' where world events, people and lifestyles are increasingly visible in our everyday surroundings. Appadurai (1996) likewise suggests that the prodigious mobility of people and goods is producing new global landscapes that we now take for granted: the French anthropologist Marc Augé (1995) offers the examples of airport departure lounges, railway terminals, 'edge-city' retail warehouses, sports stadiums and shopping malls. Each of these settings tends to be decorated in a similar manner, have identical shops and food outlets, and be enveloped by the same piped Muzak irrespective of international context. Inevitably, they will be populated by people dressed in similar ways (jeans, designer T-shirts, sportswear for the young, business suits and casual wear for older people), reading internationally syndicated magazines and comics or pausing to use their mobile phones. For Augé, these settings extend capitalist consumer values into the far corners of the globe, simultaneously

World air travel continues to grow

Revenue passenger kilometres, billions

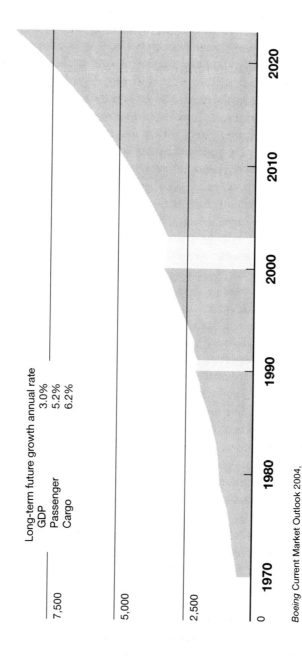

Long-term future growth annual rate

GDP	3.0%
Passenger	5.2%
Cargo	6.2%

Boeing Current Market Outlook 2004, Demand for Air Travel

Figure 5.1 Growth in global airspace movements

symbolising the sense of speed, efficiency and *supermodernity* that lies at the heart of global capitalism.

Augé has also concluded these global landscapes amount to 'non-places' – environments dominated by a devotion to mobility to the exclusion of any sense of fixity, place or local identity (Augé 1995). Such environments are often *branded spaces*, and restaurant franchises (such as global chains like Starbucks, McDonalds and Subway), airport terminals, sports stadia and shopping malls often incorporate copyrighted designs. Accordingly, they are governed by explicitly 'extra-architectural' and non-local rules of play or economic interaction (Shields 1997) and are serially produced commodities. A good example is the globally recognised branded hotels which cluster near international airports (e.g. Ibis, Travel Lodge, Comfort Inns, Sofitels). Similar in design throughout the world, and offering a standard repertoire of services, these hotels throng with travellers from around the world, all of whom typically subscribe to the global uniform of business travel – the suit. Though cynics would dismiss these hotels as depressingly placeless, Yakhlef (2004) argues they have of late gained increasing popularity because they cater to the emerging demands of globally mobile people: Bauman's (2001) 'new cosmopolitans'. As such, branded hotels now outperform unbranded, independent properties (arguably mirroring trends in global markets whereby powerful global brands such as CocaCola, Microsoft and Marlboro have undermined and supplanted 'indigenous' products).

Yet at the same time the global movement of people and goods is stretching social relations, it is evident that we are also in an era of *space-time compression*, whereby places seem to be getting closer to one another (what Marx once referred to as the 'annihilation of space by time'). Crucial here is the putative speeding-up of communication, and the increasing emphasis placed on 'weightless travel' – the rapid circulation of data, knowledge and ideas. Bogard (2002) thus describes the increasing importance of global communication networks and cybernetic systems in contemporary economic practice. As he contends, given that the contemporary global economy is characterised by an increasing flow of goods, services, capital, ideas and to a more limited extent, migrationary labour, information and communications technologies (ICTs) are vital to economic success. At the same time, ICTs have also accounted for growing shares of world trade, investment and consumption (Freeman and Louca 2002). In the 1990s, trade in ICTs grew at almost twice the rate of total trade, which itself almost doubled over that period, and technological standards became increasingly globally integrated (OECD 2002).

Consequently, *friction of distance* has been effectively reduced through a combination of technological innovations and transport improvements which, according to Harvey (1989b), occurred in two main phases. The first was at the end of the nineteenth century, as new 'enabling' technologies like the telegraph, wireless, telephone, train and steam ship permitted speedier flows of goods and faster communications (see Schivelbusch 1986). While this led to an immediate rise in international trade, it was the technological advances of the 1980s and 1990s that are regarded as having effected a truly *global* communications revolution. At the heart of this revolution was the development of the Internet and the World Wide Web, an 'Infobahn' allowing virtually instantaneous transmission of text, pictures and sounds across the world (see Kitchin 1998; Dodge and Kitchin 2000). At the same time, many people now possess mobile telephones which allow them to contact people around the world at a fraction of the cost that their parents' generation would have expected to have paid. Indeed, the overall volume of international telephone calls rose from 12.7 billion call minutes in 1982 to a staggering 154 billion in 2001 (cited in Vertovec 2004).

In consequence, while nineteenth-century urban mobility was registered in the construction of dense webs of tramlines, suburban railways and trolley-car routes, in the *wired city*, the increasing density of fibre-optic and broadband networks are testament to the importance of cyberspace communication in organising the city. Furthermore, while ICTs are undoubtedly transforming the 'internal' social geographies of the city (tying selected neighbourhoods into a network of e-retailing, homeworking and online banking), it is also significant in the reforging of intercity relations. Maps of cyberspace thus provide some clues to the increasingly commonplace transmission of ideas and information between cities on a global scale – albeit these flows are often highly centralised, with data on the location of Internet hosts suggesting that many poorer areas of the world are only weakly connected to the global networks of cyberspace.

In spite of differential access to the Internet, the increasingly important role of ICT in creating an instantaneous communication network in which traditional understandings of 'near' and 'far' are problematised is widely acknowledged. Bogard (2002: 28) alerts us to this when he stresses that 'cyberspace communications, in a word, are strange – at the push of a button, territories dissolve, oppositions of distant and close, motion and stasis, inside and out, collapse; identities are marginalized and simulated, and collectivities lose their borders.' Like many other commentators, Bogard suggests that communication innovations such as the Internet are responsible for a radical reorganisation of space-time, and more particularly, the creation of a world where national boundaries are less of an impediment to communication. Hence, while it was the cultural theorist Marshall McLuhan (1959) who first spoke of the coming of a 'global village', it is the recent 'revolution' in ICT and mobile technologies that has allowed this hypothesised village to become a reality.

Consequently, it is now often suggested that we live in a world where momentous world events are beamed 'live' into our homes via satellite technology; where products and ideas from abroad are routinely incorporated into our lives; where the political decisions that shape our destinies are as likely to be debated overseas as in our own government and where many people expect to work outside their country of birth for long periods (D. Clark 1996). In short, it seems we are living in a world where interactions of all types have been extended over space to the extent that formerly autonomous nations and locales have been drawn into contact with one another.

Given this, it has been argued that social scientists' traditional concern with the fixed and sedentary needs to be superseded by a preoccupation with the mobile and fluid. Zygmunt Bauman (2000) writes of liquidity as the defining characteristic of contemporary society: liquids may not bind or unite, but are extraordinarily mobile. These liquids ooze, seep and flow around the world, often spilling over the 'dams' and 'defences' designed to impede their progress (such as immigration controls and border tariffs). The use of hydraulic metaphors implies a need for theories that are able to make sense of these new 'geographies of flow': a point underlined by Urry (2000) when he argues for a new mobilities paradigm (see Box 5.1).

While Urry's call for researchers to elucidate the complex topologies of global travel and mobility has been widely taken up, Short (2004) rightly notes that many accounts of globalisation frequently ignore the bodily dimensions of global mobility. Thus, while we are learning more and more about the quantitative expansion of global travel and communication, its qualitative and experiential dimensions remain largely

---

Box 5.1 THE NEW MOBILITIES PARADIGM

John Urry (2004) asserts that

> there is a 'mobility turn' spreading into and transforming the social sciences, transcending the dichotomy between transport research and social research, putting the social into travel and connecting different forms of transport with the complex patterns of social experience conducted through various communications at-a-distance.
>
> (Urry 2004: 1)

Urry refers to this as a 'new mobilities paradigm'. The implications of this paradigm are apparent in many different aspects of research into transportation and movement, so that, for example, the automobile not only is increasingly theorised as a technology for moving people from A to B, but also is recognised for its role in the making of new social practices, formations and networks (including, crucially, cities). Arguing that social scientists tend to focus on the sedentary and fixed, he presents an argument for exploring the imbrication of culture and technology in practices of movement, with an explicit focus on the social spaces that orchestrate and organise different forms of physical and virtual mobility. Suggesting that mobility is not just about things and people moving *between* fixed places, there is a case made here for addressing the complex networks and topologies linking the immobile and mobile (i.e. the connections between air travel and the infrastructure of airports, or mobile phones and phone masts). This notion that mobility involves the co-evolution of technological and social systems, and the creation of self-organising systems, clearly owes much to the perspectives of Actor Network Theory (see Chapter 4), as well as non-representational theories that focus on social practices (see Chapter 3). It also serves to problematise the idea that the world is always and inevitably 'speeding up' for everyone, and argues for a nuanced understanding of the production and consumption of different mobilities. Urry's argument

continued

for a mobile sociology is a persuasive one, not least because it promises to open up the 'black box' of travel. Nonetheless, Urry's claim that social scientists have largely overlooked the social and cultural dimensions of travel is perhaps overstated, especially in the context of globalisation debates, where much work has explored the *production* of global webs of travel, communication and mobility. However, Urry is right to note that many of these are remarkably disembodied, and say little of the differential experiences of travel and communication.

Further reading: Urry (2001); Sheller and Urry (2006)

unexplored. For instance, key questions about the differences between communicating by webcam and talking face-to-face remain unanswered, despite the vast amounts spent on business travel (but see Lassen 2006). There also remains little research on the experience of flying first class rather than business class, despite the introduction of biometric fast-tracking for elite passengers which allows them to sidestep conventional border controls (S. Graham 2005; see also Cresswell 2001). In contrast, much has been written about the impact of new communication and transport technologies on *perceptions* of space and time, particularly the forging of a global consciousness. Elaborating, Urry (2000) argues that the global communications revolution has produced a montage or collage effect, so our consciousness is a mix of near and far, close and distant. This blurred experience of space has its corollary, Urry argues, in the emergence of *instantaneous time*. This includes a number of characteristics distinguishing it from the routinised (and *national*) clock time associated with industrialisation and modernisation (taken from Urry 2000: 129):

- The increasing importance of global time.
- The increasing disposability of products, places and images.
- The growing volatility of fashion, style, products, ideas and images.
- The breakdown of distinctions of day and night (and week and weekend, home and work, and so on) as new technologies change the way we live and work.

- The idea of a twenty-four-hour society (hence the emphasis on evening economies or the importance of round-the-clock financial trading).
- The diminishing importance of family rituals (such as mealtimes).
- The growth of 'just-in-time' production, so that workers are expected to be more flexible in their working hours (and companies lay claim to their whole life, not just the period they are paid to work).
- The sense that the pace of life is speeding up and has perhaps got too fast (producing notions of risk).

Far from being a banal restatement of the belief that the world is speeding up, the characteristics identified by Urry are suggestive of a two-speed society where we either move in step with global time or are in danger of being left behind. The subtext here is that we need to create faster cities to keep up in these global times. As we shall see, the idea that the city needs to become more flexible and mobile has become something of a mantra among city governors – one with important implications for urban policy.

CITIES IN THE WORLD

Summarising the links between globalisation and the city, Short and Kim (1999: 9) propose that 'globalization takes place in cities and cities embody and reflect globalization.' However, while the intensification of global processes seems to be providing a new context for urbanisation, the idea that cities have global reach and influence is not a new one. The term 'world city' was first coined by Patrick Geddes in 1915 to describe those cities in which a 'disproportionate' amount of the world's trade was carried out. Subsequent commentators on world cities – notably Peter Hall (1966), John Friedmann (1986) and Saskia Sassen (1991) – have reworked this definition in various ways, alerting to the global power of a select cadre of cities (London, New York and Tokyo) which act as global command and control centres. Beyond this, it is postulated that there is a second tier of cities which serve as major regional centres in key globalisation arenas (e.g. Los Angeles in North America, Paris and Frankfurt in Europe). A third tier is deemed to comprise international cities of lesser importance, such as Seoul, Madrid, Sydney or Singapore. Beyond this, it is suggested that there is a fourth level of nationally important cities with some strong international links (e.g., San Francisco, Osaka, Milan). Knox (1995: 239) adds a fifth tier of places like Rochester, NY, Columbus, OH or the 'technopolises' of Japan, where 'an imaginative and aggressive leadership has sought to carve out distinctive niches in the global market place'.

The basis of such world city rosters has typically been the economic attributes of particular cities for which data are readily available to researchers in the urban West, such as the number of corporate headquarters, stock exchange activity, presence of international banks, percentage of workforce employed by advanced producer services, and so on (Friedmann and Wolff 1982). While such indicators apparently reveal the most powerful and influential world cities – rather than those which merely boast a large population – we need to be mindful of the limits of such statistics, which prioritise particular forms of work over others, and generally ignore social, political and cultural dimensions of urban life (see Short et al. 1996). The fact that world cities research reifies Eurocentric conceptions of power (equating this with financial prowess) also raises serious questions about the hierarchies implied in world city rosters (Robinson 2002; McCann 2004b). Nevertheless, for Sassen (1991), understanding the making of world cities demands serious consideration of the work of *advanced producer services*, defined as those corporate activities that enable the capitalist world economy to operate through the expert assistance they give to both public and private corporations. Covering such activities as banking, accountancy, insurance, advertising, public relations, law and management consultancy, these services went 'global' at the same time as their main clients, the transnational corporation (TNC). However, the globalisation of advanced producer services has involved more than the 'global servicing' of TNCs, with many advanced producer services becoming TNCs in their own right. The most obvious example here is the banking sector (which actually includes some of the biggest TNCs). In turn, these banks are served by other advanced producer services such as law partnerships. These advanced producer service TNCs may have begun their global strategy by following clients to world cities but their continued success has required more proactive strategies. Global service corporations have thus been adept at producing their own commodities, including new financial products; new advertising packages; new forms of multi-jurisdictional law and so on (see Beaverstock et al. 2000b).

The one thing that all of these firms share is dependence on specialised knowledge. Their state-of-the-art commodities are produced by bringing together different forms of expertise to meet the specific needs of clients. In order to be able to put together such packages, it is understood that firms need to be embedded in knowledge-rich environments. Sassen (1996) suggests that world cities provide such environments, with face-to-face contacts between experts facilitated by the clustering together of knowledge-rich personnel in cities. As such, it is implied these world cities are becoming more, rather than less, important in a global era, and although the rise of the Pacific Rim means that some cities including Singapore, Vancouver or Los Angeles have taken on a new strategic importance, London, New York and Tokyo have unquestionably maintained their pre-eminence as financial centres (Taylor 2001). Ultimately, this highlights the tension that exists between the deterritorialising centrifugal impulses of information technology (which threatens to undermine place identities) and the reterritorialising centripetal tendencies that create a highly differentiated world of workplaces (where some places are more profitable and powerful than others).

Sassen's (1991) idea that the clustering of knowledge-rich individuals in London, Tokyo and New York provides a critical mass of know-how crucial to global economic success is thus central to many accounts of world city formation. Reflexivity and personal networking are also at the heart of Thrift's (1994, 1995, 1996b) understanding of the global economy, with knowledge-rich workers, firms and institutions clustering to enable the transfer of embodied knowledge. While new technologies have the potential to dematerialise these 'epistemic communities', making it possible for them to be located anywhere provided workers are electronically connected, Thrift emphasises a countervailing propensity for new forms of global control and financial reflexivity to have emerged and clustered together in particular localities. Likewise, Knox (1995) argues that it is world cities housing the 'knowledge elite' which exact the economic reflexivity that has become crucial for economic success. Knox (1995: 236) therefore identifies world cities as 'nodal points that function as control centres for the interdependent skein of material, financial and cultural flows which, together, support and sustain globalisation.' In a similar manner, Storper (1997: 222) refers to world cities as 'privileged sites', suggesting they are critical places in the perpetuation of global capitalism.

In sum, world cities are deemed to contain the economic, cultural and institutional infrastructure that channels national and provincial resources into the global economy and that transmits the impulses of globalisation back to national and provincial centres. However, such conceptions of world city formation do not challenge the idea that cities are *places* crucial to the articulation and reproduction of global capitalism. A somewhat different perspective has been offered by Manuel Castells, whose multivolume *The Information Age* (1996, 1997, 1998b) identifies a topology of flow, connectivity, networks and nodes as fundamental to global informational capitalism. Arguing that global networks are becoming the new organising and managing principles (or 'social morphology') of capitalism, he contends recent advances in information technology have provided the material infrastructure for the diffusion of the networking principle throughout society.

In so far as the logic of networks is diffusing into all spheres of human activity, Castells contends that it is also radically modifying the production, consumption and experience of space. Until the 1970s, according to Castells, modern society was constituted as 'spaces of places', such as nations, cities and regions. Contemporary network society, on the other hand, is constituted as a *space of flows*, an entangled skein of linkages, connections and relations across space:

> The space of flows refers to the technological and organizational possibility of organizing the simultaneity of social practices without geographical contiguity. Most dominant functions in our societies (financial markets, transnational production networks, media systems etc.) are organized around the space of flows . . . However, the space of flows does include a territorial dimension, as it requires a technological infrastructure that operates from certain locations, and as it connects functions and people located in specific places.
>
> (Castells 2000: 14)

Castells describes this 'space of flows' as having three levels: the infrastructural, the organisational and the social. The first refers to the hard and soft technology that enables global communication, the second to the hubs and nodes allowing the network to operate, and the third to the networks of global elites that are of paramount importance in the information age. All three have distinctive geographies, with presence or absence in these networks being 'critical sources of domination and change in our society' (Castells 1996: 469).

Elucidating the morphology of this space of flows, Castells identifies prominent world cities (e.g. London, Tokyo, New York) as key organisational nodes (places where there is also an increasing divide between an informational elite and what Castells terms the 'fourth world' – those residents in informational cities who are cut off from global networks of prestige, power and wealth – see Box 5.2). Castells in fact spends much time discussing the relationships between the infrastructural layer and this hierarchical layer of hubs and nodes, directing little attention to the third layer in a space of flows: the dominant managerial elites who have become the key players in the global economy. The neglect of this third layer has led some to suggest that Castells offers a relatively thin account of identity formation (e.g. van Dijk 1999; Bendle 2002). Nonetheless, by hypothesising the existence of three coterminous networks, Castells (2000) powerfully demonstrates that space is the 'material support of time-sharing social practices', underpinning an era where global (timeless) time exists in tension with chronological time – and a space of flows exists in tension with a bounded space of places. In this sense, the space of flows is evident in the aforementioned corporate hotels, convention centres, airport lounges and internationally serialised shopping districts which are key nodal points in the life of transnational workers and business travellers. Drawing obvious parallels with Relph's (1976) work on placelessness, and Augé's (1995) notion of non-places, Castells argues that there are now many spaces which are solely associated with the accelerated flow of people and goods around the world and do not act as localised sites for the celebration of 'real' cultures: Castells thus identifies the gradual erosion of place identity occurring as more and more sites are given over to the articulation of global flows.

Castells' rather simplistic opposition of global space and local place has left him open to criticism, particularly when his work is contrasted with the more nuanced takes on place identity offered by geographers. Exemplary here is the work of David Harvey (1989b, 1996), whose discussion of the condition of postmodernity considers how processes of time-space compression simultaneously encourage the homogenisation and differentiation of place identity. In doing so, he points out the contradictory manner in which notions of place are becoming more, rather than less, important in the period of globalisation, stressing that the

Box 5.2 THE INFORMATIONAL CITY

An influential writer in many fields of research, Manuel Castells' early work in the field of urban sociology and neo-Marxism emphasised the importance of space and helped to reorient urban geography (see Chapter 1). For example, *The Urban Question* (1977) and *City, Class and Power* (1978) drew on Marxist theory to assert that urban space was a product of capitalist industrialisation and was ultimately shaped by class and interest-group politics. Developing such perspectives, *The Informational City* (1989) and the acclaimed trilogy *The Information Society* (Castells 1996, 1997a, 1997b) described how the 'revolution' in information technologies had profoundly transformed 'the way we work, produce, consume, communicate, travel, think, enjoy, make war and peace, give birth and die' (Castells 1996: 122). Although Castells is at pains to stress that technology does not determine society, and that there is a complex interaction between technology, economic strategies and social interests, some commentators have suggested that his work still offers an assuredly structural reading, albeit that he now posits technology as the leading structure (see Merrifield 2002). In his account of the informational city, Castells stresses that a reliance on information processing and communication technologies *produces* cities which have distinctive socio-spatial forms. Most notably, he has described the informational city as a dual or *schizophrenic* city, with prestigious knowledge-rich enclaves cheek-by-jowl with pockets of poverty, dereliction and (technological) ignorance. Suggesting that these devalued spaces are bypassed by the information highways of the space of flows, he nonetheless remains optimistic about the opportunities presented by the information age to forge new socio-spatial orders, and provoke an urban renaissance.

Further reading: Castells (1996, 2001)

alleged specificity of place (in terms of its history, culture, environment and so on) is crucial in attracting mobile subjects (and mobile capital) to specific urban centres. Doreen Massey's (1991a: 145) attempt to elaborate a 'progressive sense of place' also suggests Castells' distinction of space/place is too simple. She contends 'what gives a place its specificity is not some long internalized history but the fact that it is constructed out of a constellation of social relations, meeting and weaving together at a particular locus'. Developing an *extroverted* sense of place, the implication here is that the idea of bounded places is nonsensical: they have always been open to the world (and part of a space of flows).

Notwithstanding these criticisms, Castells' structural take on globalisation has been widely influential in the literature on world cities. Specifically, Castells (1996) argues that a network of world cities is required to organise and reproduce informational capitalism, and that this network requires the connection of major world cities (e.g. London, New York and Tokyo) with cities that serve as regional hubs (e.g. Frankfurt, Miami and Singapore). In so doing, he stresses that the world city network no longer plays the same role in the articulation of global flows that was the case in the 1970s and 1980s. Then, the majority of corporate headquarters were located in major Western cities; now, new communications technologies do not require headquarter functions to be carried out only in the largest cities. In addition, the 1990s witnessed the rising importance of capital flows within so-called 'emerging markets', bypassing the corporate headquarters of the urban West. In response, TNC headquarters have often decentralised, seeking new and cheaper locations away from world city centres (which are characterised by high rents). In relation to this, it is interesting that London is home to only three of the world's largest one hundred TNC headquarters (Short and Kim 1999), but is still commonly referred to as a world city. Rather than acting as home to corporate headquarters, its role appears as a centre of *translation* vital to the articulation of global networks (that is to say, it provides the accounting, legal and business expertise which is required to maintain global businesses).

Peter Taylor (2004) has sought to reconcile and clarify Castells' work in relation to his own interests on world cities, modernities and world systems, drawing heavily on Immanuel Wallerstein's (1979) world systems theory. Taylor observes that Castells' conceptualisation of major world cities as 'command and control centres' is quite conventional in terms of the world cities literature, concurring, for example, with Sassen's (1991) identification of London, New York and Tokyo existing at the apex of a global urban hierarchy. According to Taylor (1999b), the principal contribution of Castells to the world cities literature is instead to position the world city network in a richer and more comprehensive theoretical context, making it one important network within a space of flows. Identifying world cities as the most direct illustration of hubs and nodes in the new spatial logic of the informational age allows Castells to emphasise the uneven articulation of global flow, which he reads as 'critical for the distribution of wealth and power in the world' (Taylor 1999b: 3).

Though Taylor agrees with this summation, he remains concerned by Castells' (1996) apparent conflation of the concepts of world city and mega-city (Castells describes the emergence of mega-cities as 'the most important transformation of urban forms world-wide' and 'the power centres of the new spatial form/process of the information age: the space of flows' – see Castells 1996: 403–410). This usage of the term mega-cities by Castells overlaps considerably with his deployment of the term global cities. For Taylor, these two conceptions of contemporary large cities have a very different provenance, economic-functional (in the case of world cities) and demographic-statistical (in the case of mega-cities): in simple terms, mega-cities are large cities, not all of which are global cities. As Short (2004) details, some very large cities do not appear at all significant in the articulation of informational capitalism (and he lists Theran, Dhaka, Chongqing and Khartoum as large cities which are particularly poorly connected to the networks of advanced informational capitalism).

However, Taylor (2004) makes the point that Castells' conception of global cities as 'embedded flow processes' rather than places usefully directs attention away from what cities contain (the case study approach) to their connections with other cities (the relational approach). It is this approach that informs the rosters of world cities generated by Taylor and members of the Globalisation and World City (GaWC) network. Initially focusing on the role of world cities in supporting financial and business services (which Taylor regards as particularly central to contemporary patterns of economic change), GaWC has explored how leading service firms maintain their 'brand credibility' by providing clients with a seamless service wherever the business takes them (so, for example, what begins as a London law firm becomes a successful global service firm by building a network of offices in very many world cities). In this sense, London's service cluster may be considered as a node within a world city network in which the flows are the information, knowledge, instructions, personnel and ideas that move between the offices of particular firms. Taylor (2004: 29) illustrates this by referring to the business advertising of major advanced producer service firms: for instance, Morgan Stanley Dean Witter's bold slogan is 'Network the World'; Concert promise a 'global network with seamless connectivity'; UBS Walburg boast a 'seamless global network . . . of 13,000 open minds'; HSBC draw attention to the fact they have 'over 5,000 offices world-wide'.

According to Taylor (2004), to empirically study this world city network requires data on the global location strategies of firms (i.e. where in the world they have their offices, what functions the offices have, and how important the offices are). GaWC has completed this analysis for global service firms in different business sectors (where global service firms are defined as firms with offices in at least fifteen different cities including at least one in each of the main 'globalisation arenas' – North America, Pacific Asia and Western Europe). Where a city has no branch of a particular firm it is scored zero; where it is home to a company's headquarters, it scores five; regional headquarters score four, and so on. This means that service firms are conceived as the 'interlockers' of cities as nodes, with global network connectivity of cities computed as the total sum for all firms of the summed products for firms' offices in a city with their offices in all other cities. In Figure 5.2, these are converted to proportions of the highest scoring city (London), which is the most connected (with a score of 1.0). This analysis suggests that London has the highest level of connectivity closely followed by New York. This measure can be disaggregated to consider just one particular service: for example, using just the twenty-three banking/finance firms identifies the most significant international financial centres. Disaggregating in this way suggests that Tokyo is not part of a 'big three' as a world city but does tend towards this status as an international financial centre; Los Angeles falls out of the top ten cities, while Frankfurt moves appreciably in the other direction. Pacific Asian cities are also higher and better represented in the banking connectivity list.

Calculations of the service business links for individual cities and sectors therefore begin to identify the specific roles which individual cities play within an interlocking world city network. Whereas cities have

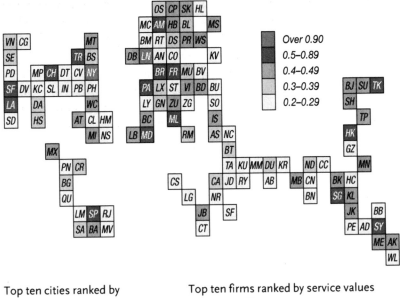

Over 0.90		■
0.5–0.89		■
0.4–0.49		▨
0.3–0.39		▨
0.2–0.29		□

Top ten cities ranked by connectivity

London	1.000
New York	0.976
Hong Kong	0.707
Paris	0.699
Tokyo	0.691
Singapore	0.645
Chicago	0.616
Milan	0.604
Los Angeles	0.600
Toronto	0.595

Top ten firms ranked by service values across 315 cities

KPMG Accountants	618
PriceWaterhouseCoopers	559
Arthur Andersen Accountants	392
CitiGroup Banking	377
Moores Rowland Accountants	367
HLB Worldwide Accountants	357
BBDO Worldwide Advertising	351
RSM Accountants	346
HSBC Bank	345
PFK International Accountants	341

Figure 5.2 Global network connectivity of advanced producer service firms (after Taylor 2004)

This cartogram places cities in their approximate relative geographical positions. The codes for cities are: AB Abu Dhabi; AD Adelaide; AK Auckland; AM Amsterdam; AN Antwerp; AS Athens; AT Atlanta; BA Buenos Aires; BB Brisbane; BC Barcelona; BD Budapest; BG Bogotá; BJ Beijing; BK Bangkok; BL Berlin; BM Birmingham; BN Bangalore; BR Brussels; BS Boston; BT Beirut; BU Bucharest; BV Bratislava; CA Cairo; CC Calcutta; CG Calgary; CH Chicago; CL Charlotte; CN Chennai; CO Cologne; CP Copenhagen; CR Caracas; CS Casablanca; CT Cape Town; CV Cleveland; DA Dallas; DB Dublin; DS Düsseldorf;

DT Detroit; DU Dubai; DV Denver; FR Frankfurt; GN Geneva; GZ Guangzhou; HB Hamburg; HC Ho Chi Minh City; HK Hong Kong; HL Helsinki; HM Hamilton (Bermuda); HS Houston; IN Indianapolis; IS Istanbul; JB Johannesburg; JD Jeddah; JK Jakarta; KC Kansas City; KL Kuala Lumpur; KR Karachi; KU Kuwait; KV Kiev; LA Los Angeles; LB Lisbon; LG Lagos; LM Lima; LN London; LX Luxembourg; LY Lyons; MB Mumbai; MC Manchester; MD Madrid; ME Melbourne; MI Miami; ML Milan; MM Manama; MN Manila; MP Minneapolis; MS Moscow; MT Montreal; MU Munich; MV Montevideo; MX Mexico City; NC Nicosia; ND New Delhi; NR Nairobi; NS Nassau; NY New York; OS Oslo; PA Paris; PB Pittsburg; PD Portland; PE Perth; PH Philadelphia; PN Panama City; PR Prague; QU Quito; RJ Rio de Janeiro; RM Rome; RT Rotterdam; RY Riyadh; SA Santiago; SD San Diego; SE Seattle; SF San Francisco; SG Singapore; SH Shanghai; SK Stockholm; SL St Louis; SO Sofia; SP São Paulo; ST Stuttgart; SU Seoul; SY Sydney; TA Tel Aviv; TK Tokyo; TP Taipei; TR Toronto; VI Vienna; VN Vancouver; WC Washington DC; WL Wellington; WS Warsaw; ZG Zagreb; ZU Zurich

traditionally been conceived as possessing bounded hinterlands in which they dominated, this shows that under conditions of contemporary global-isation servicing is worldwide (hence cities have 'global hinterlands'). Yet these global hinterlands are not homogeneous or uniform: Frankfurt and London, for example, both have strong Pacific Asian links, yet it is clear that London's is important in terms of servicing the United States, while Frankfurt has a more European focus (Taylor 2001). This empirical specification highlights that London and Frankfurt have different roles as financial and business service centres within a world city network. Moreover, it begins to suggest that cities may connect to the world in different ways by virtue of the type of firms, workers, goods and services that pass through them.

Though intuitively attractive, it should be remembered that GaWC rosters are based on a particular understanding of how the global economy works. Further, while the emphasis is on flow, certain assumptions are made about the connectivity between branches of offices within global firms: there is little consideration of actual volumes of intercity com-munication and data transmission. However, there are some measures which offer more direct indication of the world city network. For instance, global air travel data have been used as another source of information about intercity connection (see Smith and Timberlake 1995). While some measures merely indicate number of flights rather than the number of passenger movements, Derudder et al. (2004) employ the MIDT database on global airline bookings and connections relating to more than half a billion passengers to detail international air passenger flows. On the regional level, they observe that the largest flows are situated within North America and Europe, with intra-North American and intra-European flows each capturing around 24 per cent of the total amount of passengers travelling between 290 selected cities (see also Derudder and Witlox 2005). London appears the most important city in terms of arrivals and departures, attracting about 5 per cent of all the passengers travelling between the selected 290 cities. The next important city is New York, which represents 4.5 per cent of incoming and outgoing passengers. After London, the most important European cities (which capture more then 5 million passengers) are Paris, Frankfurt, Amsterdam, Rome and Milan. Following New York, the most important American cities are Los Angeles, Chicago, Washington, San Francisco and Miami. The only two (Pacific) Asian cities in the same category are Hong Kong and Bangkok, which rank respectively in the sixth and fifteenth positions (see Figure 5.3).

While such mapping of intercity relations is largely exploratory in nature, and has been criticised for illustrating strong 'empiricist' tendencies (Thrift 1998), it does allow for the specification of a world city network based on the idea of connection. Further, it demonstrates that conducting business in different parts of the world requires the support of advanced producer services, as Sassen (1991) emphasises, but that the

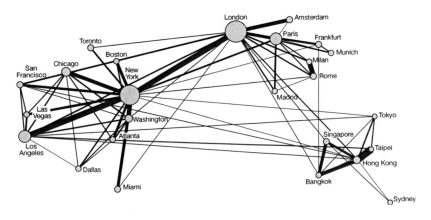

Figure 5.3 Interconnectivity of the world's twenty-five most important air hubs (after Derudder and Witlox 2005)

links and connections do not have to have a hierarchical structure. As Taylor (2004) details, for instance, developing a business project in Manchester that involves working with people in Brisbane (for example), there is no need to route the intercity connection via London at one end and via Sydney at the other:

> Messages involving information, knowledge, strategy, planning, instruction, can and may be directly relayed between the two cities in the business operation. On the other hand, the Manchester people may consider that to expedite matters they need a level of legal expertise that can only be found in London law firms. Thus all inter-jurisdictional contract negotiations between Manchester and Brisbane will go via London. The point is that with the enabling technologies of IT old hierarchies can be by-passed but they can also be reinforced. In this new global space of flows we may search out hierarchical tendencies within *some* inter-city relations but for others there will be *direct* links between cities irrespective of their relative rankings. The result is likely to be very intricate patterns when studied across many world cities.
>
> (Taylor 2004: 54, original emphases)

Taylor (2004) argues there is a key implication that derives from this argument: namely, that it is misleading to identify a small subset of 'global cities' if this is interpreted as meaning cities in which global processes are concentrated. He insists instead that under conditions of contemporary globalisation *all* cities are global: they operate in a contemporary space of flows which enables them to have a global reach when circumstances demand such connections. Hence, while many urban geographers continue to use the term 'global city' for those leading metropolises that articulate global flow, work of this type suggests it is no longer possible to talk of 'non-global' cities. In this respect, it is interesting to speculate why many urban geographers continue to do 'world cities' research given that the term implies there is a set of cities which are manifestly not world cities. McCann (2004b) certainly feels that it is vital to study and theorise 'non-global' cities to make sense of the urbanisation–globalisation nexus, stressing that an undue focus on exemplary cases distracts from the diverse urbanisms existing in both West and non-West. Similarly, Robinson (2002) suggests that making a distinction between 'world cities' and 'ordinary' cities is not just methodologically suspect, but politically dubious – not least because it perpetuates developmentalist thinking. In her estimation,

> a view of the world of cities emerges where millions of people and hundreds of people are dropped off the map of much research in urban studies to service one particular and very restricted view of significance or (ir)relevance to certain sections of the global economy.
>
> (Robinson 2002: 535)

MARKETING THE CITY AS A GLOBAL PRODUCT

Notwithstanding the criticism that city hierarchies and rosters are often highly Eurocentric and developmentalist, they have proved highly significant in policy terms as they encourage city governors to aspire to world city status. Furthermore, the idea that there are winners and losers in the battle for global jobs and investment is one that has fuelled intense rivalry between different cities. According to Harvey (1989a), this intensification of intercity competition has led directly to the rise of *urban entrepreneurialism*. Jessop (1998) lists three defining features of entrepreneurial cities:

- An entrepreneurial city pursues innovative strategies intended to maintain or enhance its economic competitiveness vis-à-vis other cities and economic spaces.
- These strategies are real and reflexive. They are not 'as if' strategies but are more or less explicitly formulated and pursued in an active, entrepreneurial fashion.
- The promoters of entrepreneurial cities adopt an entrepreneurial discourse, narrate their cities as entrepreneurial, and market them as entrepreneurial (see also Chapter 2).

The motif of entrepreneurialism captures the increasingly 'businesslike' manner in which city governments operate, taking on characteristics once distinctive to the private sector – 'risk-taking, inventiveness, promotion and profit-motivation' (Hubbard and Hall 1998: 2). Fundamental is that city governments are paying increased attention to economic production, aggressively promoting investment within metropolitan areas (Loftman and Nevin 2003). For such reasons, entrepreneurial urban governments are considered to pay more attention to the needs of the business community, and less to state provision of social and welfare services.

While the entrepreneurial strategies designed to boost urban competitiveness are many and varied (Kipfer and Keil 2002), we can identify at least two key ways in which entrepreneurial cities seek to improve their asset base so as to be more attractive to global investors. The first is to seek new 'location-specific' advantages for producing goods/services – for instance, through the installation of new physical, social and cybernetic infrastructures designed to promote agglomeration economies of scale and scope. The second is to find new sources of supply to enhance global competitive advantage. Examples here include reskilling the extant workforce or, more commonly, seeking to attract highly skilled knowledge-rich workers – those whom Richard Florida (2002b) terms the 'creative classes'. Both strategies thus embody a neoliberal ethos given that they mobilise state power in the interests of encouraging private sector investment (Peck 2001).

Extending critiques of place promotion (see Chapter 2), urban geographers have been at pains to highlight the limitations of such entrepreneurial policies for alleviating the social conditions of the least affluent, pointing out that they espouse a politics of growth rather than redistribution (see Loftman and Nevin 1998). Moreover, some have noted that entrepreneurialism is accompanied by draconian forms of social control where perceived threats to global investment are severely repressed. As we saw in Chapter 3, certain groups – squatters, homeless people, squeegee merchants and 'street people' – have no place in the spaces of the entrepreneurial city, and there is a significant connection between entrepreneurialism and the rhetoric of Zero Tolerance (Fyfe 2004). The seemingly widespread urge to tame 'urban disorder' has thus fuelled strategies of spatial exclusion justified with reference to improved quality of life, but actually intended to make the city safe for corporate gentrification (and the associated invasion of the middle classes). Despite talk of urban renaissance, Neil Smith (1996) thus portrays the entrepreneurial city centre not as an emancipatory space but as a combat zone in which global corporate capital, supported by punitive policing, battles it out, block by block, to 'retake' the city.

The alliances made between the state, police, businesses and developers suggest an entrepreneurial urban politics characteristically requires a collaboration between private and public sector actors. Deas and Giordano (2001) likewise argue that public–private partnerships are fundamental in the entrepreneurial promotion of the city because the public sector alone is unable to create, exploit, supplement and replenish city 'asset bases' and transform liabilities into assets. They contend that such actors include not only elected councillors and officials, but also planners, local business people, police, Chambers of Commerce, universities, property consortia and media interests: in short, all those who have some sort of vested interest in promoting a particular city. In the work of Logan and Molotch (1987) such public–private partnerships are termed *growth coalitions* as they represent elite groups of individuals who seek growth at almost any cost; elsewhere they are termed *urban regimes* (e.g. Stone 1989). In contrast to traditional political theories – which see power as consolidated through the ballot box – both coalition and regime theories view power as a negotiated and dispersed resource. Regime and coalition theories acknowledge that urban decision-making is diffuse, fragmented and non-hierarchical and recognise that the 'meshing of interests' brings both governmental and non-governmental actors within the ambit of governance (K. Ward 1996). Further, they insist that public–private partnerships are rarely formed from scratch, but are ongoing and constantly mutating phenomena which take different forms as urban policy evolves.

Toronto provides an example of such an entrepreneurial urban regime in action, as Tufts (2000) notes. Describing Toronto's (unsuccessful) bid to hold the Olympic Games in 2008 as a typically entrepreneurial and risky attempt to enhance the international competitiveness of the city, Tufts (2000) details that there was initially much local opposition to the idea of bidding for the Games. Clearly, the modern Olympic Games provide an important platform for place marketing as cities seek to achieve international recognition and global city status (Short 2004). However, there are often unforeseen negative effects on marginal populations, who can become more impoverished as investment is focused on accommodating the world's athletes, press and corporate sponsors. Nor is financial success guaranteed, as the public debts created by the Athens 2004 and Atlanta 1996 Games demonstrate. For such reasons, convincing local populations to support the Olympic bid was a major task confronting David Crombie, an ex-mayor of Toronto, and head of the Toronto 2008 Olympic Bid. In the spring of 1998, this committee submitted its proposal to the Canadian Olympic Association (COA), estimating the cost of hosting the games in Toronto to be 2.6 billion Canadian dollars. In February 2000, the COA and the City of Toronto formally submitted the city as a candidate to host the 2008 games. While some anti-poverty activists mobilised against this bid, they were clearly outnumbered by those who were supportive: significantly, the leaders of the bid had been careful to canvass support from a broad range of public and private sector actors, including local union representatives, convincing them that the Games would be good for the city. In contrast with the 1996 Olympic bid, which was roundly condemned by many, the 2008 bid thus received almost unanimous support (eventually just missing out to Beijing to be the host of the Games).

The way that a number of public and private institutions coalesced in Toronto around support for the Olympic bid suggests that urban regimes or coalitions may be united by the presence of a charismatic personality who is seeking to effect a particular policy or project. Conversely, the formation of a regime is necessary for this policy or project to be enacted. Ultimately, what unites actors in these regimes is the perception that a project or policy would be good for them. Hence, it has been argued that the key participants in urban regimes typically consist of locally dependent interests, whether these are the governments who want to increase their tax revenues; the utility companies who want to boost spending on gas, water and electricity; retailers who want to increase local spending or rentiers who are seeking to enhance their value of their land assets. Kevin Cox (1991) concludes that such local interests are forced to pursue entrepreneurial and outward-looking policies because they are largely *immobile* in an era where jobs and workers appear infinitely mobile.

Extending such arguments, Cox contends the relationship between social relations and spatial scale remains under-theorised in many accounts of regime formation. Cox (1993) seeks to explain the manner in which urban regimes are based within particular geographical localities by reference to the organisation of production in the capitalist world-economy. Crucial here is the idea of *local dependence*, which Cox (1993: 434) defines as 'some nonsubstitutability of local social relations'. The term describes the unique nature of particular local conditions and the consequent immobility of particular forms of capital and labour between geographical locations. In the theory of local dependency, capital production is embedded in a locality through its dependence on natural resources, physical infrastructures, labour markets and relations with suppliers and buyers found within a given territory. In turn, the labour force is dependent on a particular locality for amenities such as housing, and for social relations, including a network of family and friendship. Of course, labour may migrate, but will become equally embedded in a new city, through the process of familiarisation with its 'language' and culture, and through entry into its labour market (Cox and Mair 1989). This type of territorial constraint can also be recognised in the spatial structure of the state (for example, in the location of services such as policing and libraries, which are necessarily locally embedded).

Cox (1993) insists that local dependence continues in spite of the globalisation of the world-economy, since all capital production remains fixed within these locally dependent relations. In his view, increasing competition within the world-economy (encouraged by globalisation) leads to an increase in competition between different geographical territories, rather than individual firms. Crucially, some of these locales or cities are more competitive than others, constituting unique capital/labour complexes. Firms located within these competitive cities thus benefit from being located within their boundaries. Consequently, increasing competition from globalisation ties these firms more closely to particular localities, so that 'competition results in spatial embedding, a territorialisation of economic activity, or what was referred to earlier as local dependence' (Cox 1995: 218). The local dependency of firms means that they become 'trapped' or 'dependent' on the prosperity of a particular geographical territory. If this area experiences economic decline, the revenues that firms derive from location in this territory will decrease. In this case, there would be significant pressure for local actors to unite and form an urban regime or growth coalition intended to foster economic growth.

Through theories of local dependence, Cox specifies the conditions upon which growth-orientated urban regimes and coalitions are formed, and explains the manner in which such alliances are constructed within a particular territorial unit. Harvey (1985, 1989a) provides a similarly spatialised account of capitalist processes through an exploration of spatial flows (of money, capital, goods, credit and migrant labour) and the spatial fixities which embed capitalist production into particular locations. These fixities include the natural resources of an area, its built environment (physical infrastructure) and the spatial limits attached to the social reproduction of the workforce (i.e. the travel-to-work area of a city). Consequently, he contends that capitalist accumulation exhibits a tendency to concentrate in particular localities which become characterised by particular patterns of production and economic specialisms. However, if the generation of revenues and goods within this location exceeds demand for its commodities, or if another location becomes more competitive in the production of a particular good, there will be an outward flow of capital to other geographical areas, ensuring the creation of new 'production complexes' (and the perpetuation of capitalist accumulation).

Harvey (1989a) argues that such production complexes operate on the scale of the 'urban region', with stability within an urban region being an advantage to both the workers and capitalists who have a stake in the locality. Consequently, both have an interest in maintaining the 'structured coherence' of production systems, leading to the development of regional class alliance with the goal of stabilising processes of economic production and consumption. According to Harvey, this alliance includes home-owners, the local state, real estate developers, and the holders of business assets. However, such an alliance is prone to instability given 'the confusions of roles, orientations, and interests of individuals, groups, factions and classes' (Harvey 1989a: 152). For this reason Harvey (1989a: 152) argues that a 'relatively autonomous urban politics can arise', whereby 'politicians of genius and craft' play a crucial role in mediating between the interests of capital and labour in the construction of stable production complexes and inter-class alliances.

In a similar manner to Cox (1993), Harvey (1985, 1989a, 1996) depicts the construction of urban regimes as resting on a politically mediated alliance between actors with a dependence on a spatially defined locality. In both cases, collaboration between actors is linked to their embeddedness within a bounded geographical location, namely an urban region. Furthermore, both authors rest their analysis on an examination of the relationship between spatial scale and social relations, suggesting that particular social relations (including relations of production and social reproduction) are constituted within specific geographical scales. Consequently, actors within these locations may hold common interests in economic growth within these areas, based on their common embeddedness in these social relations. As Harvey (1996: 298) argues, cities offer a form of permanence in the flow of space and time, and it is the tension between this 'place-based fixity and the spatial mobility of capital' that he argues entrepreneurial governance seeks to resolve through 'investment in consumption spectacles, the selling of images of places, competition over the definition of cultural and symbolic capital and the revival of vernacular traditions.'

Such ideas about scale dependence are intuitively attractive, and allow us to make connections between contemporary urban politics and the increasingly globalised organisation of capitalism (Cox 1993: 433). Consequently, they exercise a considerable influence on both academics and policy-makers, informing a multitude of local economic development strategies designed to attract and embed 'footloose' capital within city regions (typically defined as the 'travel-to-work' area of a given city). Yet many of the assumptions on which such policies rest remain unproven. For instance, the claim that capital can flow 'at the twinkle of an eye from one fragmented place to another as the Starship Enterprise moves from one end of the Galaxy to another' (Swyngedouw 1989: 1) needs to be tempered with the observation that much capital remains embedded in property and investment. Nor are all workers dependent on a given locality for their housing and social needs, with transnationalism a reality for many. Moreover, MacLeod (1999) accuses theories of local dependency of buying into a naive notion of scale whereby dependency is mapped unproblematically onto the local. As he details, dependencies may exist at other scales, and one notable trend is for class alliances or networks of political association to extend across space and time to include actors and institutions whose 'stake in place' takes different forms. In particular, the fact that urban governance is being rescaled suggests that a reassessment of local dependency theory is overdue (see also MacLeod and Goodwin 1999).

UPWARDS AND ONWARDS? RESCALING THE CITY

Though most entrepreneurial policies seek to promote urban growth by manipulating the assets that are contained within the city, both theory and practice is converging around the idea that the most effective scale of territorial competition is often larger than established units of city government because these units seldom correspond to economically functional areas (Cheshire 1999). Consequently, one increasingly important strategy for enhancing urban competitiveness is upscaling the city. This refers to the process by which cities seek to redefine their spatiality, normally by working collectively with other localities towards a common end. According to Harding (1994: 39), such cooperation reflects the fact that 'city municipalities have realised, if reluctantly, that not all of the problems which their areas face are soluble within the city'. Coordination between local authorities is accordingly emerging in areas such as urban infrastructure provision and environmental regulation where the costs and benefits of services and government regulation fall across more than one municipal jurisdiction. Yet there is also increased evidence of economic development projects that make use of a pooling of resources and expertise between municipalities for their implementation. Through such mutual cooperation, city governors may seek to transcend the boundaries of urban space, possibly 'jumping scales' (so that, for example, a city with a regional profile becomes part of a more significant national or international urban conurbation, and more competitive and attractive in global terms). Such rescaling provides evidence that urban governance needs to be understood not only in terms of networks of association within cities, but also in terms of urban policy-makers' relations with other actors, spaces and, indeed, geographical scales. As we will see, this demands consideration of the ways in which influential 'urban' actors ranging from growth coalitions to individual city mayors draw on resources and networks 'above' the urban scale to promote their urban places.

In broad terms, it is possible to identify three distinct types of upscaling: the multi-tiering of government, the polycentric linking of urban functions and the creation of complex urban networks (see Savitch and Kantor 2002). The first of these implies a process by which localities can band together to establish an umbrella institution with metropolitan-wide functions. Under this arrangement, extant localities may retain a number of existing functions, but another level of government may take over many of its policy-making functions. The umbrella tier can be independently elected or separately appointed: either way it provides an institutional layer that typically supplants or supplements the bureaucratic responsibilities of lower-tier urban governments. For example, in the British context it is significant that successive Labour governments have embarked on a programme of constitutional change since 1997 that has involved the establishment of the Greater London Authority (GLA) with an elected mayor who has responsibility for the London Plan (also known as the Spatial Development Strategy for London). This covers topics ranging from strategic housing targets and locations, green belts, economic development and policies for improving movement and accessibility. As such, the GLA provides a strategic framework to guide the work of the thirty-two London boroughs, effectively supplanting the strategic planning functions of the boroughs (see also MacLeod and Goodwin 1999 on New Labour attempts to generate governance capacity through the provision of new statutory agencies for London). Beyond the capital, Regional Development Agencies and Regional Chambers (as well as the creation of Assemblies in Wales and Northern Ireland and the establishment of the Scottish Parliament) suggest that this process works in two directions, with the nation state 'scaling down' as cities 'scale up'.

In part, this 'new regionalism' is based on an understanding that the city region can operate as a competitive spatial unit. Theoretically, it is fuelled by some 'quite convincing claims' that the region has re-emerged to challenge the taken-for-granted position of the nation state as the 'pre-eminent site and scale for territorial economic organisation in contemporary capitalism' (Ward and Jonas 2004: 2119). This contrasts with the rhetoric of earlier times when city regions were seen as cumbersome and uncompetitive units, which capital and investment bypassed as it sought to locate in either the nation state or the individual firm. 'Scaling up' by becoming part of a larger city region is thus a way that municipalities might seek to enhance their competitiveness, benefiting from the agglomeration economies that exist at a regional scale:

> For certain cities, those located broadly in the same economic region (if not the same country), it can be economically advantageous to develop collaborative arrangements, networks, alliances, resources, etc with each other . . . This enables the more fortunate city-regions to specialise around clusters of economic activity, exploit comparative advantages, and outcompete cities located in other economic regions, especially those places which are seen to lack the requisite advantages and institutional capacities, or that . . . are not 'institutionally thick'.
>
> (Ward and Jonas 2004: 2122)

Of course, being subsumed within a larger city region might involve some loss of local identity, yet many urban governors see these losses as compensated by the economic benefits of being part of a larger – and more competitive – unit. Significantly, it is worth noting that many of those who make important investment decisions for large transnational firms start by looking at cities with at least 1 million population, since they consider this the minimum population needed to support the type of services and infrastructure they require. This type of logic encouraged French planners to create eight 'regional capitals' – Lyon/St-Etienne, Marseille, Bordeaux, Lille/Roubaix/Tourcoing, Toulouse, Strasbourg, Nantes and Metz/Nancy – each of which typically captured two or three urban centres within a wider metropolitan boundary. For example, the boundary of the RUL (Région Urbaine de Lyon), otherwise known as the L'Aire Urbaine de Lyon, has created a '30 mile metropolitan area around Lyon [which] contains all the facilities and infrastructures of a large urban region of European dimensions' (ADERLY 1997: 3). This scale effectively bequeaths Lyon a heightened visibility and legitimacy, its population of over 1 million symbolicly qualifies it, according to Lyon's political leaders, as a 'European metropole' and thus accords it important international status.

However, it is important to stress that countervailing tendencies also exist, with some cities having 'downscaled'. For example, Greater Manchester Council (GMC) was formed in 1974 through the amalgamation of central Manchester and nine towns in the North West of England (Bolton, Oldham, Salford, Wigan, Ashton-under-Lyne, Trafford, Stockport and Rochdale). However, in the absence of clearly defined and demarcated responsibilities between GMC and the district councils within its jurisdiction, local politicians and officers were left to negotiate these for themselves. *Streamlining the Cities*, a White Paper published in 1983, suggested that abolition would remove such conflicts, insisting that the metropolitan counties were a wasteful and unnecessary tier of government in any case (Department of Environment 1983). Underpinned by a logic of enterprise and *laissez-faire* flexibilism, this left the ten towns that used to cooperate as a city region to fight their own competitive battles (Randels and Dicken 2004: 2012). Hence, while London regional government was abolished in 1986, only to be re-established in the 1990s to produce an expanded 'unit of territorial competition', the towns and cities of Greater Manchester have moved in the other direction. As Randels and Dicken (2004) imply, this is because central Manchester already considers itself as a competitive urban region, and feels it does not need to collaborate with Bolton or Rochdale to compete with continental European cities such as Lyon and Milan. In this context, it is interesting to note that Manchester is involved in a number of domestic (e.g. UK Core Cities) and international networks (e.g. Eurocities) which do not include any of the other towns in the Greater Manchester area.

This brings us to a second form of rescaling, which involves the linking of specific functions across a number of towns and cities within a given region. These linked functions usually depend upon an agreement being made about the specialisms which might be best provided in specific locales. For instance, the French government encourages this form of functional differentiation through a system of contracts with individual cities (*contrats de villes*). These contracts may advance a particular industry (*technopole*) or rejuvenate neighbourhoods or adopt strategies for combating social exclusion. In the aforementioned example of Lyon, for example, the discourse of Le Reseau de Villes is that each city will promote its particular sector strengths: Lyon in health and the environment, Grenoble in electronics, St-Etienne in architecture and the built environment, and so on. Similarly, in the English East Midlands, the Three Cities Town Net project is seeking to address the region's 'as yet unfulfilled potential to become the region of choice for domestic and overseas investment', building on Leicester, Nottingham and Derby's specific specialisms (such as Leicester's leadership in space research, Nottingham's excellence in biotechnology and science and Derby's tradition of aeronautical engineering). This initiative is jointly funded by Leicestershire Promotions, Nottingham City Council, Derby City Council, the European Commission, the Office of the Deputy Prime Minister, Nottingham East Midlands Airport and the East Midlands Development Agency (EMDA), in keeping with EMDA's aim for the region to become one of Europe's top twenty performers by 2010. Again, it is significant that each of the cities involved has a population of around 250,000: by collaborating, they can claim to have over 1 million population within their travel-to-work area, meaning they are able to compete with more visible and better known cities across the world.

This form of upscaling ties in with the identification of *polycentric urban regions* (Box 5.3) as globally competitive spatial units par excellence. These are defined as regions with two or more historically and politically separate cities which are nonetheless functionally interconnected. A notable example here might be the urban network formed by the 'Flemish Diamond', which takes the form of a set of physically separate urban areas. The three largest cities – Brussels, Antwerp, Ghent – and the regional city of Leuven form the anchors of this diamond-shaped urban constellation,

Box 5.3 POLYCENTRIC URBAN REGIONS

Although it has a long precedent, the idea of polycentric urban regions is particularly associated with the geographer-planner, Sir Peter Hall. His argument is that the high-level control functions of powerful global cities (such as London, Paris or New York) are dispersed across a wide geographical area, limited only by certain geographical constraints of time-distance. As he describes, 'though traditional face-to-face locations retain their power, they are increasingly supplemented by new kinds of node for face-to-face activity. The resultant geographical structure . . . is quintessentially polycentric' (P. Hall 2001: 73). Such polycentric urban regions are characterised by separate and distinct urban nodes which interact with each other to a significant extent. European examples which are frequently mentioned include Randstad in the Netherlands (Kloosterman and Lambregts 2001), the Flemish Diamond in Belgium (Albrechts and Lievois 2004), the Rhine-Ruhr in Germany and Padua-Treviso-Venice in northern Italy. Outside Europe, the southern Californian sunbelt in the United States and the Kanasai area in Japan have been identified as polycentric regions. In this respect, the notion of polycentricity is clearly connected to the concept of *megalopolis* – a term first coined by Jean Gottmann (1987: 2) to describe the US eastern seaboard cities (Boston/New York/ Washington – or BOSNYWASH) – 'a chain of national and international crossroads . . . which owes its destiny to a web of far-flung and multiple networks of linkages with the whole world' (see Baigent 2004 on the work of Gottmann). Despite the widespread use of terms such as megalopolis, the conceptualisation of polycentricity remains characterised by a lack of clear definition of the concept, unconsolidated literature and little agreement on criteria and thresholds for analysis (Kloosterman and Musterd 2001).

Further reading: Kloosterman and Musterd (2001)

which is home to more than 3.3 million people. Like the more familiar examples of the Randstad and Rhein-Ruhr, the Flemish Diamond is cited as an example of a polynuclear urban region as its centres are not merely physically proximate but are also functionally integrated, in the sense that the cities within the complex perform different functions. For example, governmental activities are concentrated in Brussels while port activities and industry are concentrated in Antwerp (Albrechts and Lievois 2004).

However, the definition of a polycentric urban region remains problematic in at least two respects (Parr 2004). First, there is the question of how the functional interdependence between cities can be measured and adjudged to indicate connectedness. While labour market flows are often taken as an adequate measure of interdependence, there seems to be a lack of evidence of interactions between some of the most frequently cited examples of polycentric regions. For instance, Lambooy (1998: 457) suggests that 'as yet there is no strong evidence that the Randstad [Hague-Amsterdam-Rotterdam] area may be considered as one city region', suggesting that the area is merely a 'geographical image on the map' rather than a functional reality. Musterd and van Zelm (2001) similarly assert that the Randstadt is less a series of interdependent cities and more a set of adjoining city regions. Second, there is the issue of what constitutes spatial proximity, Here, most commentators have suggested travel between any two points in a polycentric urban region should not take more than a 'comfortable commuting time'. One hour's travelling time is often taken as a maximum (Bailey and Turok 2001), although other limits for travel time have been employed (see Davoudi 2002). Although the question of how long an acceptable commute might be is debatable, the introduction of rapid-transport systems connecting the different nodes of a polycentric urban region provides the possibility that such regions might become more dispersed. On this basis, some suggest that many of the established polycentric urban regions of North Western Europe (e.g. the Randstad, the Flemish Diamond and the Rhine-Ruhr Area) might even be thought of as 'metropolitan macro regions'. However, this also raises the question of the minimum separation between centres, and again there is little agreement on this (Parr 2004). Without some degree of separation, however, it has been contended polycentric urban regions would include instances where urban centres have coalesced to form a conurbation (e.g. Greater Glasgow, Lille-Roubaix-Tourcoing and Minneapolis-St Paul).

Despite the question marks remaining about the definition and conceptualisation of polycentric urban regions (PUR), considerable stress has been laid on the supposed economic advantages of the PUR, particularly in terms of its capacity to foster cooperation and to permit the efficient exchange of goods, services and information. Nonetheless, Parr (2004) argues it is very difficult to accept that these advantages are unique to the polycentric urban regions and are therefore not present in economic systems based on alternative spatial structures, particularly in an age of continually improving transportation and telecommunications systems. Somewhat related to this is the focusing of attention on the internal networking of the polycentric urban regions, to the neglect of external links. Given the current concern about the competitiveness of areas, Parr concludes more attention should be paid to the *international* links of the polycentric urban region and the firms located within it, particularly with respect to the sources of inputs and the location of markets.

Hence, although many planners regard polycentric urban regions as offering a model for sustainable urban development and capitalist growth, the creation of local intercity alliances is potentially as myopic as the strategies of urban entrepreneurialism pursued since the 1980s. Indeed, the idea that international competitiveness may be best enhanced through bringing together cities and towns which are already well connected is something of an oxymoron. Nonetheless, this idea is widespread in the literature on urban and regional development, where urban competitiveness is frequently understood to be a function of the assets contained within a specific territory (see Figure 5.4). Developing the work of management guru Michael Porter, Kresl (1995), for example, is adamant that external aspects must be excluded from any analysis of the determinants of a city's competitiveness. Likewise, he insists that a city's international competitiveness is quite different from the concept of an international city, arguing that the latter concerns connectivity (at the international scale), while the former concerns only the city in question. In his view, a city can be extremely competitive without being connected into a network of other cities, just as a city can be fully plugged into a network without being at all competitive. It is on this basis that Kresl (1995) claims that a city may dramatically increase its competitiveness – even its international competitiveness – without being or increasing the degree to which it is an international city.

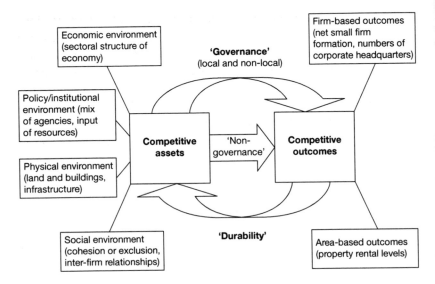

Figure 5.4 Urban assets and competitive outcomes (after Deas and Giordano 2001)

For Kresl, city competitiveness and success unquestionably are a function of the city's asset base. Likewise, in the influential work of Krugman, which dismisses the notion of competition and holds to the idea of strategic economic complementarity, urban success is deemed to derive from agglomeration economies at the local scale (Boddy 1999). Against this, Phelps (2004) insists that *agglomeration economies* now appear to be available over much less localised geographical scales than the city regions classically identified as competitive economic territories. His argument is that the growth of diffuse urban forms suggests a need to move away from classical agglomeration theory and the tendency to simply upscale the territory over which external economies are perceived to be available. Parr (2002) concurs, detailing the sources of competitive advantage associated with particular forms of external economy:

External economies of *scale* . . . are based on the existence of specialised servicing activities, the possibilities for cooperative research and development activity, and the advantage of industry-wide marketing. By contrast, external economies of *scope* refer to cost savings to the individual firm which are dependent on the existence of firms in other industries. Here the economies are based on the shared use of inputs with other firms . . . A third externality is concerned with external economies of *complexity* . . . The cost savings in this case derive from efficient information flows and the ability to coordinate its activities with other firms, particularly with respect to the avoidance of input-supply problems. Each of these external economies may exist in a variety of spatial settings, *including ones in which the relevant activities have a dispersed pattern of location.*

(Parr 2002: 721, emphasis added)

Developing this argument, Parr (2002) details the type of economic advantage which may accrue to a firm locating in a given city region (such as access to public utilities, transportation services and other infrastructure, as well as specialised business services). Yet he also notes there are spatial diseconomies associated with locating in cities themselves, and that many of the benefits of being located in a metropolitan region can actually be enjoyed at a distance. This is particularly the case in the service sector, where economies of scale may be less important than in the case of manufacturing. As Coe and Townsend's (1998) examination of new firm formation and clustering in the United Kingdom makes clear, localised agglomeration economies make little sense in relation to the service sector, where many firms enjoy the advantages of access to London's global markets even though they may be highly decentralised (located in cities throughout the 'Greater' South East, including Peterborough, Hemel Hempstead and Milton Keynes).

Debunking the myth that agglomeration economies are created only through physical proximity, contemporary work in the 'new economic geography' suggests more attention needs to be devoted to the networks of interdependence and synergy created between firms which may be located at a distance. Phelps (2004) thus argues that the use of existing theory by economists and geographers which emphasises external economies of scale provides an increasingly inadequate means of understanding contemporary forms of urban economic agglomeration. His suggestion is that more attention is paid to network *externalities*, noting there are many instances where firms agglomerate for reasons that are nothing to do with the presence of other firms. The upshot here is that extra-regional linkages may be more important than inter-regional linkages.

If one stops thinking of agglomeration economies as localised clusters with tightly circumscribed boundaries and starts thinking of spaces of economic flow, the notion of city competition changes. Here, the success of a city is no longer reducible to its locally embedded resources, firms and infrastructure, but concerns the inputs and outputs that move through it. To talk of these flows benefiting one city to the detriment of all others is clearly nonsensical: urban economies do not primarily compete with each other since there needs to be collaboration and a division of labour for the global space economy to 'work'. Sassen (1991) illustrates this with reference to the financial deregulation of the 1980s. This occurred at a time when the relations among the major financial centres – New York, London and Tokyo – were seen as one of straight competition (so that the 1986 Financial Deregulation Act, for example, was supposed to enhance the competitiveness of the City of London). Sassen (1991) suggests that deregulation favoured no one city over the others, but enabled all three to enhance their position in the world-city network. Likewise, New York's domination of investment banking and fund management may well have been considerably more important than growth within the UK space economy for strengthening London's position in the world-city network.

In spite of these observations, many examples of urban upscaling appear to be based on a rather straightforward conception of international competitiveness, whereby a city region can outcompete its rivals by offering a distinctive and significant spatial agglomeration of know-how and institutional capacity within a given territory. In contrast, a third option for rescaling lies in the 'complex network' approach. Here, no tier of government is added of any substantial nature, and there is no attempt to redraw the boundaries of the city. Rather, officials of existing independent localities make decisions to cooperate through multiple, overlapping agreements. This idea emphasises the importance of synergy, allowing localities to trade on each other's strengths. In effect, networking of this type emphasises that 'cities cannot make it on their own in an era of globalisation and increased interurban competition, but need to cooperate' (Leitner and Sheppard 2002: 495).

What is perhaps most significant here is that these complex municipal networks may be spatially diffused, involving cities which are not physically proximate. For example, the Lyon-Marseille 'Co-operation Charter' was agreed in 1997, bringing these distanced cities together to 'reinforce . . . the strengths of each of these metropoles, cornerstones in the new territory of prosperity of the Greater South-East of Europe' (Lyon Cite,

January 1998). Underpinning this was Marseille's wish to regenerate its port activities, and Lyon's wish to renovate and extend its inland port terminal, taking freight off congested roads. Other networks may be cross-national, building on traditions of town-twinning (see Vion 2002). An example is the 'Transmanche Metropole', a 'cooperative initiative involving strategy between the city authorities of Caen, le Havre and Rouen in France and the English local authorities of Southampton, Portsmouth, Bournemouth and Poole' (Church and Reid 1996: 1307). This network aims to share knowledge between city governments on approaches to economic development, with Church and Reid (1996) explaining that these areas face a unique type of regional economic problem based on restructuring ports, depressed resort towns and a rural hinterland with marginal activities. Likewise, the Town Net project is one of the transnational projects in the European Union's Interreg IIIB programme for the North Sea area. In this project seven participating city regions, from The Netherlands, United Kingdom, Norway, Sweden, Denmark and Germany, have teamed up to carry out a wide range of activities falling within four themes: networks and linkages; spatial quality; economic cooperation and regional/town-network identity. The project's main objective is to strengthen the participating regions as a whole by stimulating development of each town's specialism, encouraging greater cooperation between towns, cities and regions without competitive overlap and resulting in differentiation and complementarity.

These examples suggest that policy-makers are beginning to embrace a politics of cooperation where the intention is to enhance connectivity between cities. In many ways, these strategies acknowledge the economic realities of a world where global networks provide the dominant social morphology. However, it is surely dangerous to argue that work beyond the confines of the city should be prioritised over work within. More prosaically, perhaps, there seems to be a case for academics and policy-makers alike to question inherited understandings of competition and agglomeration, inside and outside, local and global, near and far. Entrepreneurial politics, polycentric urban networking, regional alliances and transnational networks may all play a role in enhancing city competitiveness, but what seems to be badly lacking at the moment is any monitoring of how these strategies succeed in establishing new flows between the cities – which is crucial if cities are to be more competitive in these globalised times.

CONCLUSION

In this chapter, I have begun to explore the ways that geographers have tried to make sense of the city in an era where established notions of scale are being undone. As I have suggested, such attempts at reconceptualising the spatiality of the city have been driven by the realisation that global flows are deepening and thickening exponentially. For some, the new-found hyper-mobility of (urban) capital, migrants, workers and commodities unroots the city from time and space, to the extent that they find it difficult to imagine cities as bounded space-times with definite surroundings (Crang 2002). For others, this intensification of mobility and movement merely signals a change in the role of cities as articulators of capitalism, with some cities gaining, and others losing, key roles as staging points in global networks. Hence, from some perspectives, cities remain crucial nodes in global networks, while from others, cities are global networks in and of themselves. Overall though, there seems to be an emerging consensus that it is important to do away with the idea that cities are contained within a global world that shapes their destiny.

There is thus an emerging literature attuned to the urban geographies of flow and mobility. Even so, the idea that cities have an 'inside' and 'outside' remains predominant, as does the idea that cities are fixed locales in time and space. As we have seen throughout this chapter, dominant conceptualisations of cities as places remain stubbornly recalcitrant, even in the face of rampant globalisation. As Amin (2004) notes,

> it will take a lot to displace the A–Z or concentric-circle image of London by a *relational* map that incorporates the network of sites around the world that pump fresh food into a distribution centre called Covent Garden, that draws neighbourhood boundaries around settlements in postcolonial countries with which social and kinship ties remain strong, that makes us see sites such as Heathrow airport or Kings Cross station as radiations of trails shooting out across the land and far beyond.
>
> (Amin 2004: 34)

Yet perhaps this relationality is what is truly interesting about cities. Accordingly, geographers are slowly learning that it is important not merely to map a city's fixed position within an increasingly interconnected global network, but also to explore the way it stretches through time and space. Rather than thinking about cities as islands of economic competitiveness or territories of social reproduction, perhaps we need to think about them as a tangle of flows which drift into and drift out of a multitude of spaces and times, speeding up and slowing down, contracting and expanding as they are shaped by human and non-human actants. It is this type of argument that Smith (2003: 577) highlights when he argues that 'city networks are best understood not through points, lines, and boundaries (or the language of clustering, agglomeration, and localization) but as . . . continuous circulations of flux and chaosmosis where there can be no summation and so no integrity'. The implications of this form of post-structural thinking for the conceptualisation of cities are profound: working through these issues thus represents a major challenge for contemporary urban geography.

FURTHER READING

There are a number of good, accessible introductions to debates surrounding cities under conditions of contemporary globalisation. David Clark's (1996) Urban World, Global City is broad in scope; Short and Kim's (1999) Globalization and the City is more focused, and provides a good introduction to debates around place marketing and globalisation. The Entrepreneurial City (Hall and Hubbard 1998) is an edited collection which considers the implications of global 'place wars' for city governance. Saskia Sassen's (1991) The Global City stands as perhaps the most important discussion of world cities as centres of economic influence, while Peter Taylor's (2004) World City Network provides a detailed overview of a series of related mappings of the networks created by global firms. The Open University reader Unsettling Cities (edited by Allen et al. 2001) also contains a number of very accessible chapters on the connectivity of cities in a global era. Finally, it is important to stress that much of the literature on world cities is based on Western and Eurocentric notions of competition and finance: for a critique, see Jenny Robinson's (2006) Ordinary Cities: Between Modernity and Development.

6

THE CREATIVE CITY

Cities have always played a privileged role as centres of cultural and
economic activity. From their earliest origins, cities have exhibited
a conspicuous capacity both to generate culture in the form of art, ideas,
styles and attitudes, and to induce high levels of economic innovation and
growth.

(Scott 1996: 323)

This book began by suggesting that urban theory had run out of steam by
the 1990s. While urban questions remained of interest to geographers,
few sought to contribute to urban theory *per se*, or made any definitive
claims as to the distinctive nature of urban space. In contradistinction
to rural studies – where interest in the definition of the rural rejuvenated
the discipline – urban studies became a rather more diffuse endeavour
as questions of what constitutes cityness were sidestepped. In Chapter 1,
I elaborated on the background to this state of affairs by presenting a brief
genealogy of urban theory. This overview identified some of the most
widely celebrated explanations of the form and function of cities. Though
clearly selective, the conclusion drawn was that it is hard to discern any
sense of progress over 150 years of urban studies. In sum, it appeared
urban geographers remained fixated on serving up reheated versions
of the same old accounts of city life. This was illustrated with reference to
the shared emphases of modern and postmodern urban theories (e.g.

their identification of the city as restless, complex and prone to bouts of creative destruction). As such, both postmodern and modern cities have been explained away with reference to the organisation of the capitalist political economy, 'natural' processes of social sorting (underpinned by bid-rent), and people's desire to counter the sense of loss they experience in the city by forming self-sustaining communities.

Suggesting that many of these explanations are logical, elegant and, in their own terms, entirely plausible, Chapter 1 nonetheless noted recent scepticism about their ability to make sense of cities as distinctive spatial formations. One upshot is that urban geography appeared somewhat moribund in the 1990s, lagging behind other areas of the discipline in terms of its theoretical ambition and sophistication. Yet there are grounds to suppose that geographers have now circumvented what Thrift (1993) termed the 'urban impasse', and are beginning to develop new understandings of the city. Some notable examples have formed the basis of Chapters 2–5 of this book. To recapitulate: the previous chapters have turned away from some of the more established (and arguably tired) theories of cities to consider approaches which take its materiality more seriously. In simple terms, these four approaches propose the following:

- that representations of the city carry material weight, with the symbolic forms and languages of city space entering into the creation of its (so-called) 'real' spaces
- that city spaces are created and transformed through material practices which are, in many cases, beyond representation
- that urban spaces are material networks enrolling a multiplicity of actants, where the latter includes both human and non-human actors
- that urban spaces are more or less 'stretched-out', and have a material existence that problematises extant ideas of geographical scales.

Although there are some tensions and contradictions between these different approaches, my argument in this chapter is that they shed new light on the concept of the city by alerting us to its material forms. Taken together, these perspectives suggest that the city is a unique and profligate combination of 'stuff' – buildings, practices, rituals, technologies, texts, animals, screen, networks, objects and people. Arguably, it is this very materiality that previous urban theory often effaced, eviscerating our appreciation of what differentiates the city from, for example, the networks and spaces of the rural.

The suggestion that urban geography needs to scrutinise the materiality of the city arguably forces us to return to unfashionable questions about the very definition of cities. And here we might do worse than return to Wirth's (1938) identification of population size, heterogeneity and density as defining characteristics of urbanity. Oddly, many urban theories make little of these, even though it is the sheer weight of people in cities that bequeaths them their distinctive character and ambience. And, if we add in the dense collection of other 'stuff' that is part of city life – objects, ideas, money – perhaps we are close to grasping the materiality of cities and understanding why their geographies remain fundamentally different to those of the rural.

This book accordingly concludes in this chapter by suggesting that urban theory will progress only by revisiting debates about the distinctive and excessive materiality of cities. Suggesting that such ideas hold for both world cities and 'ordinary' cities in both Western and non-Western contexts, the chapter nonetheless focuses upon a specific issue to demonstrate how materialist approaches may develop our understanding of cities. This issue is *urban creativity*. At time of writing, this is a key issue in Western political circles, with policy-makers seeking ways to enhance the vibrancy of cultural industries. Yet while the cultural industries are increasingly being seen as a significant source of employment within deindustrialising cities (Landry 2000), there is an ongoing debate about how creativity can be best fostered. In this chapter, I want to review some of the main ideas that abound about urban creativity, suggesting that while they offer some insights into the production of culture, they do not take the materiality of the city seriously enough. This, I will argue, results from the failure of many writers to engage with the more mundane and ordinary aspects of creative work and to address the varied ways in which city life is represented and performed. Following the structure of the book, in the remainder of this chapter I therefore take four 'cuts' through the literature on urban creativity, considering the representational, practical, networked and mobile dimensions of creativity.

In considering urban creativity in this section, I do not wish to reify creativity in any way, or suggest that it is the only route to boosting urban economic vitality. On the contrary, there may be good reasons to suppose the creative industries will not provide a long-term and sustainable form of urban growth. Indeed, much writing on urban creativity works with a restricted (and Eurocentric) model of creativity which devalues many forms of cultural production. Further, there is also a tendency to glamorise the creative dimensions of work in the cultural industries, and underplay the mundane (and sometimes poorly paid) nature of these jobs (Crewe et al. 2003). One problem here is that the majority of studies of creativity (including most cited in this chapter) are based on studies of cutting-edge creative industries in metropolitan centres, particularly world cities. This ignores all sorts of creativity that goes on in cities which is not part of the formal economy in any sense. After all, economic practice comprises a rich diversity of capitalist and non-capitalist activities and it is increasingly argued that we need to make non-capitalist ones 'visible' lest we reify capitalism (Gibson-Graham 1996). That said, the debates around creative cities are mainly concerned with the task of making money, and to that end, the next section explores the ways Western cities are imaginatively seeking to attract the 'creative classes'.

THE IN-CROWD: CREATIVE CLASSES AND THE CREATIVE CITY

In its everyday usage, creativity refers to all manner of imaginative and innovative practices. It is easy to suggest, in fact, that all forms of production and consumption require some level of creativity to be exercised (e.g. writing or reading this book!). However, in contemporary policy circles, creativity connotes a somewhat narrower sphere of concerns, centring on the ability of individuals and groups to produce artefacts or displays which have cultural or artistic merit. As such, the term 'creative industries' is often used as a shorthand to describe the convergence between the arts sector and the media and information industries, with the creative industries commonly taken to include advertising, architecture, art and antiques, film-making, designer fashion, software, music, the performing arts and television and radio. The sector thus includes many activities which have their origin in individual skill and where commodities are imbued with considerable aesthetic and symbolic value (A. Scott 2000). Nonetheless, this sector cannot be regarded as synonymous with the cultural industries, given that many commentators take the latter to include enculturated forms of production in which the outputs are not primarily cultural (A. Pratt 2004). Likewise, it is distinct from those service sector activities such as tourism, hospitality or sport, where it is an *experience* rather than a cultural good that is being produced and sold.

In post-industrial societies, where the production of tangible commodities has apparently been superseded by a dematerialised *weightless economy*, knowledge and creativity are often described as the raw materials of economic prosperity. Given this emphasis, it is widely believed that the creative industries have a significant potential for wealth and job creation. Further, creativity is encouraged as it is seen as a way of promoting cultural diversity and democratic access to culture. Consequently, many reports on national competitiveness suggest the creative industries represent a panacea for decline and disinvestment, and exist on the cutting edge of the post-industrial economy. In the United Kingdom, for example, the election of New Labour in 1997 ushered in a raft of policies intended to promote the creative industries, not least the establishment of the Creative Industries Task Force (A. Pratt 2004). Defining the creative industries as 'those industries which have their origin in individual creativity, skill and talent and which have a potential for wealth and job creation through the generation and exploitation of intellectual property', this task force estimated the contribution of the creative sector to the UK economy as 8 per cent of gross domestic product. It also suggested this sector was expanding at around 9 per cent per annum (compared with 2.8 per cent for the entire UK economy) (Department for Culture, Media and Sport (DCMS) 2001).

This identification of the creative industries as drivers of national economies means creativity has become a potent symbolic territory in the emerging twenty-first-century knowledge economy (Jayne 2005; Peck 2005). This is mirrored in a gamut of urban policies which seek to foster innovation and creativity at the local level, most notably, through the designation of cultural or creative quarters. In the United Kingdom, prominent examples include Sheffield's Cultural Industries Quarter, Stoke's Cultural Quarter and Birmingham's Media Quarter. In fact, it is not only the larger cities that have sought to boost their creative industries, with smaller cities and towns (e.g. Huddersfield and Folkestone) having developed designated cultural districts. In each case, the hope is to create a vibrant and innovative milieu in which creative industries can flourish, often benefiting from close ties with tertiary education. In this sense, almost every city is currently promoting its claims to be a 'creative city' (see also Chapter 5).

In many ways, this 'creative cities' discourse is tied into a tacit understanding that cities cultivate creativity. In his magisterial review of *Cities in Civilization*, Peter Hall (1998) powerfully underscores this argument when he identifies the city as a cultural crucible, arguing that great movements in art and culture can be located in specifically urban contexts. His suggestion – that the creative flame burns more brightly in the city than the country – is borne out in his overview of the 'golden ages of civilisation': periods when the creativity that produces both great art as well as new technologies flourishes (and here he dwells on Athens in the fifth century BC, sixteenth-century London, eighteenth- and nineteenth-century Vienna, fin-de-siècle Paris and Los Angeles in the twentieth century). In part, Hall argues that creativity is a condition of urbanity because so much of urban life is about finding creative solutions to urban problems (e.g. how to get clean water to a densely packed population, or how to promote cooperation between people of different origin and aspiration). Yet for him it is also about the concentration of wealth and power in cities – the implication being that creativity needs investment (and wealthy patrons).

Yet Hall (1998) also seeks to identify something else – that oddly intangible *essence* of cities that may, however briefly or intermittently, foster creativity. His assertion that there is an element of serendipity about the way that great cities fuel creativity is perhaps an unsatisfactory conclusion, but does suggest a launching pad for further explorations of urban creativity: in effect, Hall challenges urban geographers to explore why the chance encounters and happy accidents that fuel creativity happen in particular cities at particular times. As I will describe in this chapter, this poses important questions about how we conceptualise the city as a material space.

Of course, Hall was not the first to make the connection between great cities and civilisation, with Weber (1923: 272) having famously suggested that 'the city and only the city has produced the characteristic phenomena of art history'. This equation between culture and the city is frequently made, whereas the rural is often regarded as a backwater of cultural production, situated away from the cutting edge of creativity and innovation (see Chapter 2 on urban myths). De Vries (2001) counters, however, arguing that great art does not thrive exclusively – or even especially – in great cities. As he argues, in these days of Internet communication and hypermobility, there is no necessary reason why creative individuals need to live in cities to communicate with other creative people. Likewise, he points to works of art whose origins and sources of inspiration have not been urban, but rural. In many senses, he is right to inject this note of caution, and remind us that the connection between creativity and the city is a myth. Nevertheless, like other city myths, the idea that cities are more cultured, creative and cutting edge than the countryside has important material effects, and is a myth played on by many city governors as they seek to boost their local economy. As Hall (1998) notes, in the 1980s and 1990s, cities across Europe – Nîmes, Grenoble, Rennes, Glasgow, Cologne, Manchester, Bologna, Valencia – became increasingly occupied with marketing themselves as European cities of culture, believing that this branding would make them more attractive to investors and mobile professional workers. The title of European City of Culture has been much sought after, and widely touted as a marker of international credibility (Gold and Gold 2005).

This concern with cultivating a cultured image and ambience has become de rigeur in an era where filling the gaps left by warehouses and factories is a political priority. At time of writing, the work of Richard Florida (2002b) offers one of the most widely cited accounts of how cities acquire competitive advantage in a global economy, and is a significant influence on creative city policy. In his view, regions and cities develop advantage based on their ability to create new business ideas and commercial products. Such innovation, it is contended, depends upon the concentration of highly educated, knowledge-rich and inevitably mobile workers within a given locale. These are the creative workers who Florida alleges give city economies their cutting edge, and allow certain cities to out-perform their rivals. In Florida's work, indexes of creative talent are shown to be significantly correlated with high-technology industrial locations characterised by high rates of growth and entrepreneurship. In the United States, such locations identified by Florida (2002b) include San Diego, Washington and Boston (Table 6.1). Not incidentally, Florida notes these cities also feature prominently in lists of North America's best cities, offering a quality of life which apparently has much appeal for the creative class. Here, Florida (2002b: 7) emphasises the social-cultural dimensions of city life, suggesting that the 'creative class don't just cluster where the jobs are . . . they cluster in places that are centers of creativity and also where they like to live'.

Table 6.1 Large US cities' creativity rankings

City	Creativity index	Creative workers (%)	Creative rank	High-tech rank	Innovation rank	Diversity rank
1 San Francisco	1,057	34.8	5	1	2	1
2 Austin	1,028	36.4	4	11	3	16
3 San Diego	1,015	32.1	15	12	7	3
3 Boston	1,015	38.0	3	2	6	22
5 Seattle	1,008	32.7	9	3	12	8
6 Chapel Hill	996	38.2	2	14	4	28
7 Houston	980	32.5	10	16	16	10
8 Washington	964	38.4	1	5	30	12
9 New York	962	32.3	12	13	24	14
10 Dallas	960	30.2	23	6	17	9
10 Minneapolis	960	33.9	7	21	5	29

Source: Florida (2002a)

While Florida's (2002b) analysis stresses the importance of creative individuals in a wide variety of fields – engineering, design, management – his ideas are highly relevant to explaining why some cities are leading players in the creative industries. In effect, he argues that the nexus of competitive advantage is seen to have shifted to those regions that can generate, retain and attract the best 'talent'. The suggestion here is that the driving force behind the development and growth of cities is the ability of a city to attract and retain creative individuals. Consequently, processes of city branding and place marketing typically target the creative classes by emphasising their cultural credentials (Peck 2005). Given that Florida asserts that creative people shun the suburbs in favour of 'happening' inner city districts, this may involve the conscious marketing and commodification of urban quarters which are branded cosmopolitan and chic. An example here is the Wicker Park district of Chicago, which Lloyd (2002) characterises as an area of music and arty production that was seized upon by the national media in the 1990s as a 'cutting-edge' neighbourhood, becoming a magnet for artists, students and young professionals. In this case, while the incoming occupiers reduced the heterogeneity of the area (the majority being white), the ethnic diversity of the area appeared a major lure for artists. Lloyd (2002) suggests that some of the characteristics perceived by marketeers as liabilities are thus important in fostering cultural quarters. For instance, while Business Improvement Districts (BIDs) have been established in many US cities to 'cleanse' urban space in the name of white, middle-class consumption, these are in danger of effacing the 'messy vitality' which many artists seek out (see Steel and Symes 2005, on BIDs). Patch (2004: 175) puts it more bluntly, arguing that white artistic gentrifiers avoid 'deracinated neighbourhoods' as they regard ethnic diversity as a key contributor to local ambience.

What is perhaps most interesting about Florida's (2002a) analysis is that city diversity is seen to be a significant factor encouraging the creative classes to locate in a particular city. As he puts it, talented people report a clear preference for places with a high degree of demographic diversity, preferring places where 'anyone from any background, race, ethnicity or sexual orientation can easily plug in' (Florida 2002a: 750). In one sense, this implies that in an era where cities – like firms – compete for talent, low entry barriers to talent are a distinct advantage. Florida cites a variety of evidence to support this, noting that in societies where immigration is dissuaded, economic stagnation is the norm rather than innovation. In many ways, this chimes with one of Peter Hall's (2005) key observations about great cities – that they are open to the world:

> Plato's Athens, Michelangelo's Florence, Shakespeare's London, Mozart's Vienna. All were economic leaders, cities at the heart of vast trading empires, places in frenzied transition, magnets for talented people seeking fame and fortune. Outsiders made these places what they were: Athens's version of green-card holders, the noncitizen Metics; the Jews in 1900 Vienna; foreign artists in Paris around the same time. They were all patrons because many had made money from trade, as well as artists. They occupied a special marginal position: not at the heart of courtly or aristocratic establishments, yet not entirely shut out either. And thus they absorbed and reflected the huge tensions between conservative and radical forces that threatened to divide these societies.
>
> (P. Hall 2005: 17)

Given this, the measure of diversity that Florida employs in his empirical studies is somewhat interesting, as it is not a measure of ethnic diversity, but the percentage of a city's population that identifies as gay or lesbian. Markusen (2005: 7) finds this 'glib' measurement of diversity troubling, arguing that the positive correlation between diversity and high-tech production which Florida uncovers is actually the result of a variable which Florida omits from his analysis – educational attainment. Not withstanding this critique – as well as the sheer meaninglessness of statistics on sexual orientation – it is interesting to speculate as to why Florida equates the 'diversity' the city with its proportion of 'gay' inhabitants: as Bell and Binnie (2004) note, the stereotyping of gay, lesbian, bisexual and transsexual communities as inherently creative is problematic, as are policies designed to attract gay men and women to cities on the assumption they are childless, high income and interested in conspicuous consumption (Box 6.1).

Box 6.1 GAY VILLAGES

As with other forms of sexual practice that contradict heterosexual norms and ideals, historically there have been rules and regulations in place which have tended to discourage gay, lesbian, bisexual or transsexual individuals celebrating or displaying their identities in public. As such, they have tended to remain effectively invisible to the heterosexual majority (leading to the metaphor of the 'closet'). Yet in some instances they have become more visible, with recognised areas of gay space emerging – particularly (though not only) in cities. In the United Kingdom, for example, the gradual legalisation of homosexuality has been mirrored in the progressive formation of recognised gay 'villages', with Manchester, Birmingham, London, Brighton and Blackpool having areas recognised as hubs of gay (male) life. In many cases, these were previously marginal areas which offered affordable property for what was a marginalised population. Yet it also needs to be noted that some of these inner city areas exercised considerable appeal for young gay males, especially for those whose lifestyle revolved around city centre clubs, cafés, pubs and other facilities. Explaining the location of gay spaces is, however, complex, and many commentators have sought – with mixed success – to integrate political economy and cultural choice theories (Lauria and Knopp 1985; Brown 2001). What is perhaps significant, however, is that gay spaces are associated with a specific (and

continued

remarkably globalised) set of commodity forms, being highly aestheticised and celebrating conspicuous consumption. It is for such reasons that urban authorities are increasingly keen to promote gay quarters, believing they attract significant numbers of affluent, young and creative consumers. Yet the processes by which gay villages are created, commodified and marketed point to some key contradictions in gay space: while providing a haven for many affluent gay men, they marginalise many other groups and identities (such as lesbian women as well as older, less affluent and non-white gay men). Furthermore, it might be argued that the naming and segregation of gay spaces merely constitutes a strategy designed to enclose and 'Other' sexual minorities, reproducing the hegemonic heterosexualisation of urban space (see Hubbard 2000 on heterosexual geographies).

Further reading: Brown (2001)

Notwithstanding this, the attraction of creative talent is deemed to depend upon the ability of a city to sell itself as a cosmopolitan, creative or funky city – the type of environment where creative individuals will enjoy working. Krätke (2004) confirms this when he suggests that the extraordinary post-unification growth of Berlin as a media and culture centre is due to its appeal for the creative class as a place for living and working. Representations of the city are clearly not incidental to this process, with the idea that some cities are 'cool' endowing them with cultural capital. In turn, representations suggest that these are places where people will gain cultural capital just by being there. As Molotch (1996: 229) argues, there is a positive connection between place identity and the type of brands and firms which become associated with that place: 'favourable images create entry barriers for products from competing places.'

One example is Manchester, a city which has been aggressively marketed and branded as an oasis of metropolitan and cosmopolitan chic. This branding has relied upon the careful cultivation of a Manchester 'script' which provides a shared discourse which all politicians, agents and cultural intermediaries involved in the urban regeneration of Manchester can draw upon:

Manchester is more than a geographical location or a political entity; it is a state of mind. It is Manchester United, the birthplace of computers, railways and Rolls Royce; it is excellence in education. It is Oasis and the friendliest international airport in the world.

(*Marketing Manchester*, cited in K. Ward 2000: 1096)

The key motifs here are that of a brash self-confidence, and a belief that Manchester sport, business and music are the best in the world. It is thus notable that Manchester's marketeers rarely mention their UK rivals for investment, but have sought to position Manchester as a *European* city (comparing Manchester's café scene to that of Amsterdam, for instance). The fact that Manchester has spawned a series of notable music 'scenes' – including the 'Madchester' rave scene (centred on the Hacienda nightclub and Manchester-based Factory records acts such as Happy Mondays and New Order) – is particularly important here, signalling that Manchester is a hub of home-grown creativity in music and culture. The fact that such scenes are associated with drug use lends credibility to the boast that Manchester is tolerant and liminal, and this fuels a creative vibe. Nonetheless, the perpetuation of a 'Manchester script' relies upon a careful series of connections being made between the implied creativity of the city, its aesthetic profile and a peculiarly Mancunian attitude. Founded in 1996, the Marketing Manchester organisation has drawn upon the design and selling skills of local entrepreneurs and artists to consolidate this brand identity, working up distinctive slogans and advertisements utilising bold 'Manchester' typefaces. In the British context Manchester has thus been seen as being at the vanguard of policy initiatives based on the promotion of cultural entrepreneurialism (Binnie and Skeggs 2004).

One of the contradictions of this process is that the scripting of cities as creative draws attention to certain dimensions of difference, but effaces others. For instance, in Manchester, active attempts have been made to strategically incorporate the 'gay village' into the marketing of the city, suggesting that the cosmopolitan vibe of this regenerated area is indicative of Manchester's openness to difference. However, Binnie and Skeggs (2004) argue that the use of the gay male (in particular) to mark out cosmopolitanism has depended not only on them remaining in the position of the safe, usable Other, but also on a significant proportion being depicted as threatening. Likewise, they allege that the gay village itself is experienced as an exclusionary space by many women of colour. Here, it is notable that interviews with some of those attracted to live in Manchester's burgeoning city centre housing market suggest that they have been attracted by this image of cosmopolitanism and tolerance but appear to remain troubled by certain class and race identities (Young and Millington 2005).

WHERE IT'S AT: EXPLAINING EVERYDAY CREATIVITY

The suggestion that images of cities draw creative people to cities clearly flags up one dimension of the connection between cities and creativity. In particular, this suggests that the creation of a buzz around a particular city will encourage the clustering of like-minded creative people, creating a critical mass of cultural entrepreneurship. Yet this explanation fails to elaborate how creative individuals are stimulated by the city as a living environment. As we saw in Chapter 3, a conceptual focus on cities as lived-in places demands that we explore how individuals practise the city on an everyday and sometimes mundane level. Crewe et al. (2003) suggest these non-representational dimensions of the creative industry are lost in geographical accounts, and there is a need to situate creativity more squarely in its material and embodied contexts of production. Florida's (2002b) exploration of the rise of the creative class, for example, makes scant reference to the spaces in cities which provide the infrastructure for individuals to nurture their creativity.

The idea that particular cities inspire great art and culture is an established one, but few have sought to explore how this occurs. Molotch (2003: 95) nonetheless claims that every designer 'draws from the surrounding currents of popular and esoteric arts and modes of expression – verbal, literature and plastic – that make up everyday life'. Clearly, different surroundings provide different currents, and it is this that needs to be explored when seeking to diagnose what gives a city a creative ambience. Although it is somewhat dangerous to characterise artists as flâneurs (see Chapter 3), many claim to be inspired by the sense of vitality which exists in particular cities, where diverse encounters and experiences may trigger artistic production. Drake (2003) suggests that creative workers' personal or emotional responses to place will affect how they use the 'attributes' of that place for aesthetic inspiration, and develop a sense of artistic identity related to the locality.

By way of example here, we might consider a city currently renowned for cultural production. Although traditionally overshadowed as a centre for art and design by New York, Paris and Milan, London has emerged to become an important locus of global art and culture. At present, London's creative industries are relatively strongly clustered, with the primary cluster in the West End (edging towards Islington and Camden) and secondary clusters in west London and the City fringes to the east. It is this latter area that has become particularly renowned for modern art, with the growing market for modern art by young British artists important in the selective regeneration of Hoxton, Shoreditch and Whitechapel. Artists started moving into Hoxton during the late 1980s and early 1990s when old warehouse spaces suitable for studios were available at cheap rents. Currently, there is a cluster of around eighty commercial and non-commercial art galleries in the area, including studios associated with some of the biggest names in contemporary British art: Damien Hirst, Gavin Turk, Dinos and Jake Chapman, Rachel Whiteread, Chris Ofili, Gillian Wearing, Sarah Lucas and Tracey Emin (many of whom studied at Goldsmiths College, South London). Throughout the 1990s, these young British artists challenged the perceived elitism of the London art scene via provocative exhibitions (notably the 1997 *Sensation* show). In turn, their activities transformed the locale:

Hoxton was invented in 1993. Before that, there was only 'Oxton, a scruffy no man's land of pie and mash and cheap market-stall clothing, a place where taxi drivers of the old school were proud to have been born but were reluctant to take you to. It did not register so much as a blip on the cultural radar. Hoxton, on the other hand, became the first great art installation of the Young British Artists: an urban playground tailor-made to annoy middle England, where everyone had scruffy clothes and daft haircuts and stayed up late, and no one had a proper job. By the end of the 90s, Hoxton had spawned an entire lifestyle: the skinhead had been replaced by the fashionable 'Hoxton fin' as the area's signature haircut, the derelict warehouses turned into million-pound lofts. As the groovy district du jour, Hoxton had come to represent the cliff-face of the cutting edge, and everyone wanted a piece. At Tracey Emin's opening at the White Cube 2 gallery in Hoxton Square in 2001, Guardian art critic Adrian Searle recalls, 'You literally couldn't get into the square'.

(*Guardian*, 21 November 2003: 17)

Although the Hoxton scene has since been satirised, the transformation of the area was remarkable: galleries such as Matt's Gallery, the Chisenhale and White Cube in Hoxton Square (Plate 6.1) rapidly transformed from being marginal spaces to major landmarks on the tourist map of the city.

Plate 6.1 White Cube gallery, Hoxton Square, London (photo: author)

In seeking to understand the dynamism of the London art scene in the 1990s, While (2003) claims that extended networks of associations, co-ops, facilities and studios were essential in bringing together and promoting an international and avant-garde art movement. The idea that London is a critical nexus in global flows of art is thus an important one (to which we will return in the next section). But this argument does not explain why Hoxton and the eastern fringes of the City, in particular, became the centre of an innovative art movement. While (2003) cites Zukin (1995), who insists that 'cultural capitals' must be considered more than entrepôts of art consumption given that they are also spaces where art is made. Here, it is important to explore the 'more-than-representational' appeal of the City fringes to artists as a space in which to live and work. Like other urban locales, a neighbourhood like Hoxton is not only perceived, but also lived, experienced and practised. As has been suggested, artists and creatives might be drawn to such areas because these spaces reflect their cosmopolitan disposition and can be *rescripted* and reimagined as artistic districts with little resistance. This might also be understood as related to a process whereby creative individuals are attracted to an area by low rents, instigating a process of gentrification (Ley 2003). While the literature on gentrification helps to understand why areas like Hoxton emerged as trendy art districts in the 1990s, Latham (2003: 1704) argues that it does not provide 'an adequate path through the ordinary, intimate relations' that are central to the emergence of culturally resurgent areas.

Consequently, it is instructive to think about not only the image of Hoxton as attractive to creatives, but also the rituals of inhabitation and occupation played out there. As with areas of artistic production elsewhere (see Lloyd 2002; Latham 2003), it is notable that Hoxton and the City fringes became home to a panoply of trendy restaurants, boutiques, clubs and pubs in the 1990s. Long-established pubs such as the Bricklayer's Arms (Charlotte Street) and the Betsey Trotwood became renowned as spaces where artists mixed with media creatives, students and Britpop musicians, while new clubs and restaurants (Cargo, 333, Swish, The Electricity Showrooms, Troy, 291 Gallery etc.) sprung up around Shoreditch and Hoxton. This coincidence of artists' residences and spaces of alcohol consumption was accordingly emphasised in Martin Rowson's map of artists' London, inspired by George Cruikshank's nineteenth-century satirical cartoons (Plate 6.2).

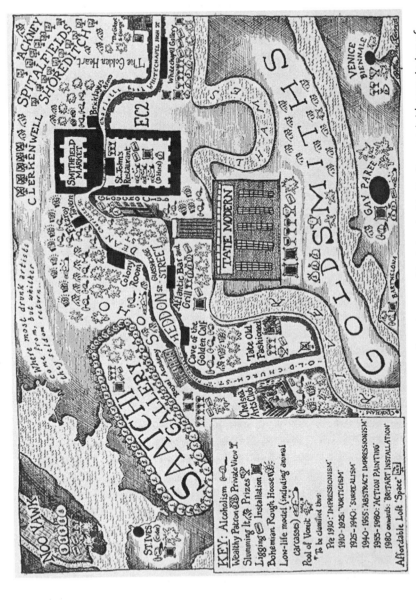

Plate 6.2 Art, alcohol and the social spaces of London: a satirical view by Martin Rowson (with permission of the artist)

Noting that there is a correlation between artists' residences and trendy bars is one thing, but exploring how artists consume these spaces is quite another, as Latham (2003) notes. The consumption of food or alcohol may be the pretext for a meeting, and gives sociability a purpose. Such forms of consumption may well be mundane, but become productive of a new type of space and new forms of sociality. As we shall see in the next section, creative work is inevitably embedded within social networks which strengthen and support what might otherwise be solitary and isolated working practices. Hence, spaces of sociality are spaces of creative work in as much as they are part of the 'field' where artists and creatives position themselves as part of the 'scene' (Grenfell and Hardy 2003). This of course extends to many other areas of creative work, where the exchange of ideas, knowledge and gossip is endemic to being on the cutting edge of innovation. This implies that the distinction between work and non-work becomes problematic in creative locales, in the sense that work becomes a fun, sociable activity where creatives are surrounded by, and interact with, like-minded friends and colleagues. Spaces of play (bars, clubs, pubs) are consequently spaces where serious work occurs. This is a point eloquently made by Hochschild (2001), who suggests that work is a source of pleasure for the many who make their work – and not their family – the centre of their social world.

The idea that creative communities congregate in spaces where consumption can become focuses of creativity is thus important in appreciating urban creativity. This is highlighted in Allen Scott's (1999, 2000) influential work on urban cultural production, which emphasises that the local atmosphere of creativity cultivated by creative individuals fuels a *collective creativity*: local cultures shape the nature of economic activity, while economic activity becomes an integral and dynamic component of local social life. This symbiosis of economy, culture and place involves more than just a coming together of people in places of sociality: there is also the question of how the public spaces of cultural production present artists with a density of sights and sounds which inspire their art and fuel their creativity. For artists, it may well be that certain localities provide the *visual* 'raw materials' which inspire them and give them a creative edge over artists located elsewhere. The local built environment may be especially important as a source of aesthetic inspiration (Drake 2003), but so too is the mix of people and objects that characterises particular locales. Latham's (2003) study of the Ponsonby Road area of Auckland (New

Zealand) suggests that artistic and creative areas are characterised by a benign tolerance of difference, and the main characteristic of such areas is the presence of a 'motley' array of people from different walks of life. In seeking to make a statement on the nature of British society, it was clearly important for Young British Artists (YBAs) to escape from the established (and elitist) spaces of the art world to an environment that seemed to be more representative of the diversity of multicultural Britain. While (2003) thus argues that Young British Art needed to be located away from the traditional and more polite environs of West London, with the East End providing a ready source of found objects and quotidian influences which YBAs incorporated in their art. In the work of Michael Landy, for example, this took very literal forms, with the rubbish and objects found on the streets of the East End becoming part of his *Scrapheap Services* installations, designed to catch the mood of social change, disaffection and redundancy wrought by the enterprise culture in working-class areas. More generally, the street life and diversity of the area has been claimed to nourish the experimental and avant-garde culture of the Young British Artists: the animation of the streets fuelled creative imagination in a number of significant ways (see also Groth and Corijn 2005).

In many accounts, art spaces are referred to as *bohemian* because of the characteristic social and aesthetic sensibility that becomes associated with them. Originally referring to the spaces of nineteenth-century art production in Paris, the label bohemia carries with it a romanticised imaginary of freewheeling, countercultural experimentation and excess (see E. Wilson 2000). This is often contrasted with the lifestyle and spaces of the bourgeoisie, who favour a more Protestant work ethic and adhere to a more respectable set of moral codes and conventions (Florida 2002c). The distinct 'structures of feeling' created in bohemian spaces thus reclaim the main elements of what Lefebvre (1991) termed 'differential' space: it is space created and dominated by its users through bodily practices which value quality over quantity, the look over the gaze and the sensual over the scientific. Bohemian zones may also be characterised as 'free zones' whose functional and economic role is difficult to explain in terms of capitalist economics (Groth and Corijn 2005). In practical terms, this allows for a wide spectrum of use which is capable of integrating a high degree of diversity. This arguably produces a form of emancipatory 'urbanity' which fosters creativity – albeit a form of creativity often highly valued by the market. The relationship between bourgeois and bohemian is consequently seen as full of ambiguity, with the former often extending their influence into the spaces of the latter so as to integrate people who think 'outside the box' into conventional business networks.

One implication is that bohemian spaces are short-lived, as they are quickly incorporated into a more conventional set of practices as the bourgeoisie move in. The outcome is often the creation of a pseudo-bohemia or *bobo* (i.e. bohemian-bourgeois) space. The media popularisation of 'edgy' artistic quarters consequently often leads to increased commercial interest in such areas and spiralling property prices which displace the 'original' creative pioneers. As Solnit (2000) laments, part of what makes a city vital and stimulating for artists – the braiding together of disparate lives and diverse cultures – is threatened by a gentrification which yanks out some of the strands altogether, diminishing urbanism itself. As she details, the Left Bank of Paris and London's Bloomsbury no longer beckon artists; bohemia has been all but driven out of Manhattan as the last pockets of poverty get gentrified, and districts like Wicker Park (Chicago), Pioneer Square (Seattle) and SOMA (South of Market, San Francisco) seem to be going the same way (Solnit 2000).

The story of Hoxton and Young British Art is one that fits into this oft-told story of urban revivification and subsequent gentrification. The story goes something like this: in the first stage, a bohemian fringe discovers a neighbourhood's special charms, such as social diversity, architectural heritage, spaces of alternative sociality (Godfrey 1988). Non-traditional 'footloose' elements are particularly numerous, comprising members of urban subcultures, artists, students and 'dropouts'. These 'urban pioneers' make a run-down or even dangerous area liveable and attractive to others who would not normally venture there – and, in so doing, they encourage the beginnings of housing speculation (N. Smith 1996). Improvement by individuals is followed by entrepreneurial investment, and ultimately corporate speculation and middle-class gentrification. The edgy character of the area is blunted as it drowns in a sea of Starbucks' café latte.

According to this academic discourse, it is tempting to suggest we need to look towards conventional urban geography explanations about property rental values to understand the origins (and demise) of creative quarters. But to return to Latham's (2003) arguments (described above), to really appreciate the way that gentrification occurs, we need to move away from such meta-theories of political economy and zoom in on the specificities of practice which collectively produce specific forms of 'embedded' gentrification. Patch (2004: 181) is particularly attuned to this, arguing that even if real estate capital tries to push people in particular directions, it does not mean it will happen. So, while exploring the profit-seeking motives of real estate capitalists provides an understanding of gentrification tendencies, this needs to be complemented with an analysis of how gentrified space is produced from the multiple interactions between the body and the city – in short, the ordinary practices of urban habitation.

CHANCE MEETING: MAKING NETWORKS OF CREATIVITY

Before we leave the example of art, it is worth working through another aspect of its creation: the importance of networks which sustain and support artistic production. As Solnit (2000) contends, the closer that artists are to museums, publishers, audiences, patrons, politicians, other enemies and each other, the better this is for them and for their art. In the case of the Young British Artists, the strength of the group and the art they made was cemented by the close relations among the exhibiting artists. This is illustrated in Figure 6.1 with reference to the social networks which centre on Damien Hirst, one of the most celebrated and infamous of the London-based artists. Significantly, this network includes artists, patrons (most notably, Charles Saatchi) and institutions, with particular seats of learning (Goldsmiths College) implicated, alongside exhibition spaces (the ICA, White Cube and Tate Modern). The social networks visible are of various duration and thickness, and include intimate relations: Michael Landy with Abigail Lane, Sarah Lucas with Gary Hume, Fiona Rae with Stephen Park. According to Grenfell and Hardy (2003), these networks facilitated the exchange of ideas and provision of mutual support that these and other artists offered each other during the 1980s and 1990s. These links persisted in the artists' choice of commercial representation, with many represented by Jay Jopling's White Cube Gallery.

This take on creativity implies that artistic creativity may be embedded in social networks which are spatially localised. The idea that creativity benefits from the dense spatial concentration of creative workers in particular locales might thus lead us to describe areas such as Hoxton as a milieu of creativity. The identification of milieux of innovation in specific spaces represents a familiar trope in economic geography, where the

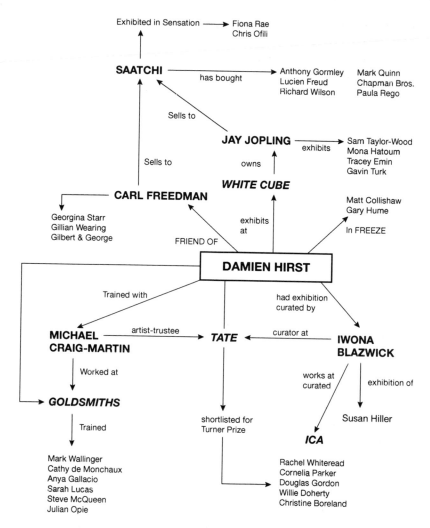

Figure 6.1 Art networks: Damien Hirst located in networks of creativity (after Grenfell and Hardy 2003)

discovery of particular 'territorial complexes' of research, design and production has been fundamental in the evolution of theories of clustering and competitive advantage. Studies of high-tech clusters in Silicon Valley, film production in Paris, the motor sport industry in the English Midlands, documentary-making and animation in Bristol, craft production in Emilia-Romagna (Italy), business innovation around Wall Street (New York) or advertising in Soho (London) have thus provided the basis for developing key ideas about why the clustering of people and firms matters (Scott 2001; Grabher 2002; Bassett et al. 2003).

The recent interest in clusters or agglomerations of economic activity has drawn intellectual nourishment from diverse analytical traditions in economics and geography. Yet, as the list of cited successes of industrial agglomeration suggests, clusters are theorised to exist at a variety of scales, from the neighbourhood to the region. Major examples of industrial agglomeration may be as small as a street or as large as the world cities that Amin and Thrift (1992) refer to as key neo-Marshallian nodes. Notwithstanding this diversity, in most instances, commentators have suggested 'embedment' of workers and firms is reproduced through two simultaneous economic and cultural processes: economic transactions between firms in the production chain, and informal social and cultural interactions between firms and consumers (or what Storper 1993 refers to as 'untraded interdependencies'). In both cases, notions of proximity are crucial. Glaeser (2000: 84) hence argues that 'cities exist to eliminate the transport costs for people, goods and ideas'. In a similar fashion, Storper and Venables (2004: 353) insist that people and economic activities tend to agglomerate to facilitate links between firms; to ease access to markets; to create dense labour markets; and to spark 'localised interactions which promote technological innovation'. Scott's (2002) writing on the clustering of French film production in Paris postulates another idea, namely the city itself provides a dense network of theatres, concert halls and other spaces where human and cultural resources combine in unexpected and new combinations to engender spontaneous *learning effects*. Rather than being 'in the air' (Santagata 2002), cultural information is generated and circulates in these sites, producing new knowledges.

Hence, ideas about inter-firm linkages, flexibilisation, the learning economy and untraded interdependence may all be invoked as explanations for the development of particular pools of expertise and talent in particular places. Yet doubts remain about the particular weight that can be attached to these different explanations, and a search for a general theory of economic clustering remains fraught with difficulty given the very different divisions of labour associated with different economic sectors. In short, while the clubs and bars of central Tokyo, New York and London might be important sites where gossip and exchange of information foster the continuing clustering of economic expertise, this type of explanation may be less relevant in the context of a science park where researchers working for different firms rarely – if ever – socialise, and where there is little inter-firm interaction.

What is widely agreed, however, is that knowledge networking is more crucial in the creative industries than in more traditional manufacturing industries, with Crewe and Beaverstock (1998) arguing that 'the importance of informally constructed social networks between firms, in the guise of sharing local intelligence, industrial-specific knowledge or trust and cooperation, cannot be underestimated in the strive for industrial competitiveness'. This is because the creative industries rely on knowledge of market trends and consumer trends to produce innovative (and often idiosyncratic) cultural goods and performances: in the creative industries, to be out of touch or off the pace is to face financial ruin. Moreover, it is clear that much knowledge in the creative industries is not explicit (in the sense that it is written down or codified) but is *tacit* (in the sense that it is embodied and practised). Creatives thus learn from others through observation, talking and experience rather than just referring to technical manuals, and this implies that there is often no substitute for meetings and teamworking. From this perspective, knowledge is deemed to be created in the 'generative dance' between actors who learn from one another by watching, talking and writing to one another (Cook and Seely-Brown 1999).

Grabher (2002: 209) thus suggests that creatives need to locate in specific locales to situate themselves in a milieu of 'rumours, impressions, recommendations, trade folklore and strategic misinformation'. His case study of advertisers in Soho (London) reveals that advertisers pride themselves on their ability to pick up on emerging cultural, social and political trends, meaning project meetings might be required at short notice. Having key connections nearby is not necessary for such meetings to occur, but is certainly convenient (Grabher 2002). One motif emerging here is that of *intensity*, and the capacity of given locales to engender multiplicity and flux (Scott 2001). As Grabher's (2001) study of advertising makes clear, the intensive networking of creatives seldom leads to the emergence of a standard set of knowledges or 'best' practices, but triggers specific forms of 'adaptation, recombination and rejection'. Further, his study of advertising creatives emphasises the project-based nature of much work in the creative sector, where individuals from different firms need to collaborate on one-off commissions. These projects constitute 'temporary social systems' in which people of different backgrounds collaborate to achieve a complex task in which learning is key. This serves to make the point that clustering is not always about long-lasting networks of the cultivation of trust among individuals, but about creating the *ability to act* (a point echoed in the communities of practice literature – see Wenger 1998).

Increasingly, theories of knowledge production in industrial clusters suggest knowledge is not a resource which is shared and distributed between firms, but is an emergent property of networking. In this sense, processes of learning and knowing are more important than processes of knowledge distribution, and tacit knowledge has precedence over explicit knowledge. Given this, it is perhaps not surprising that face-to-face contact remains central to the coordination of the economy, despite the remarkable reductions in travel costs and the astonishing rise in the complexity and variety of information – verbal, visual and symbolic – which can be communicated instantaneously (Storper and Venables 2004: 351) (see Table 6.2). Many commentators on economic clusters thus come to the conclusion that rich, thick networks of association can emerge only in contexts where face-to-face communication is possible, and even in an era where air travel has become normalised, the sheer convenience of being located in a localised cluster of like-minded creatives and firms perpetuates these agglomerations.

What the literature on clustering also suggests is that cities promote a form of networking that is seldom restricted to interactions between workers involved in the same sector of activity. Neff (2005), for instance, insists that the strongest innovation effects of co-location may actually involve actors linking across organisational and sectoral boundaries. This kind of 'networked individualism' fosters a community approach to creative production, links firms to one another through the social ties of their respective employees, and draws upon the social practices associated with 'neo-bohemians' (Lloyd 2002) to bring innovation inside the firm. For the cultural industries, location in urban locales where there is a pool of creative expertise in a variety of cultural production and distribution may thus be crucial.

By way of example, one can consider the emergence of what Power and Jansson (2004) term a 'post-industrial' economy of music production in Stockholm. They suggest that music is being increasingly distributed

Table 6.2 Advantages of face-to-face communication

Function	Advantage of face-to-face	Context
Communication technology	High frequency Rapid feedback Visual and body language	Non-codifiable technology information R&D Teaching
Trust and incentives in relationships	Detection of lying Co-presence a commitment of time	Meetings
Screening and socialising	Loss of anonymity Judging and being judged Acquisition of shared knowledge	Professional groups 'Being in the loop'
Rush and motivation	Performance as display	Presentation

Source: after Storper and Venables (2004)

electronically via information technologies, with downloading and buying compact discs (CDs) online becoming the norm and the ringtones charts an important indicator of an act's success. This digitised music production nexus nonetheless requires the support of a number of associated creative industries: web design and advertising services tailored to musical products, software programming and design focused on online products and virtual instruments; high-tech post-production and mixing services and digitalised post-production processes such as video and sound mastering. They suggest all these new activities seem dependent upon the existence of a mass consumption market for music and, to a large extent, the continued production and selling of CDs, making Stockholm an obvious site for the centralised production of Swedish music.

Less tangible assets were also commonly referred to in Power and Jansson's (2004) study, with every firm interviewed mentioning the positive influence they felt from just 'being there'. Living and working close to family and old friends, local competitors and partners, and having the chance to 'listen' to the local milieu and trying to find the 'next big thing' were all significant factors. Networks between creative workers thus evolved from personal friendships and acquaintances made by firm members during their early careers in the music industry (and to a lesser extent the ICT industry). These networks developed as 'structured' networks which linked together important gatekeepers, possible clients and collaborators; perhaps more important was that they created a placed identity for the music, creating what is referred to as a 'scene' (see Halfacree and Kitchin 1996). The dominance of Stockholm in Swedish life, the dense fabric of the city and the fact that nearly all the firms were located within walking distance from one another backed up the 'buzz' (see Box 6.2) surrounding the city itself.

Box 6.2 BUZZ CITIES

The notion of buzz refers to the all-important but ill-defined sense of a city being 'the place to be'. Although the term has been deployed in many contexts, it has been codified by Storper and Venables (2004) as the cumulative (or *superadditive*) outcome of a sequence of face-to-face interactions which promote trust and group-working between individuals in particular contexts. They contend that people in a buzz environment are highly motivated and are well placed to share knowledge and ideas with others. What is particularly important in their account is that this buzz is deemed a result of co-location rather than 'intermittent interludes' of facework. They suggest that social ties created via face-to-face meetings rather than long-range communication have a richness and durability which is important in sustaining the creativity of particular cities. Likewise, they suggest that spatial proximity fosters innovation by facilitating unplanned, haphazard inter-network meetings. This suggests that while information technologies facilitate the dis-aggregation of production, innovation depends upon on *noises* – 'rumors, impressions, recommendations, trade folklore, strategic misinformation' (Neff 2005: 134) – that can be picked up only 'on the ground' and face-to-face. Although restating the importance of face-to-face merely reiterates long-established ideas in economic geography, Storper and Venables' (2004) notion of buzz implies that many of the activities deemed central to the late capitalist economy are of a type and nature where buzz is all important (and implies that it is less important in some well-defined sectors of manufacturing and export). For policy-makers, a key question is whether it is possible to produce local social and cultural conditions amenable to the cultivation of face-to-face. Given the cumulative nature of knowledge production and apparent embeddedness of knowledge-rich individuals, it seems that it may be difficult for cities which lack buzz to becoming buzzing cities. However, this will not stop them trying, and the widespread citation of the ideas of geographers and management theorists (such as Storper (1997), Porter (1998), Florida (2002b), Krugman (1998)) illustrates the way that theories of cities become part of the policy environment (and hence enter into the constitution of the city itself).

Further reading: Storper (1997)

Power and Jansson (2004) conclude that music and ICT agglomerations in Stockholm can be described as 'clusters' since they are geographically agglomerated and exhibit intensified internal transactions; also they have well-developed institutional frameworks and cluster organisations, specialised labour forces and training/educational schemes, developed cluster brands and marketing operations, identifiable social networks, and some level of group consciousness among industry workers. However, the idea of collaboration and collective consciousness is only part of the story, as clustering of this type might also fuel 'healthy' rivalry. This is emphasised by Malmberg and Maskell (2002), who suggest:

> We suspect that the reason for the relatively meagre results coming out of studies attempting to study empirically the magnitude and intensity of local interfirm collaboration may be that it is on the horizontal dimension of interfirm observation, comparison, and rivalry that the effects of agglomeration are most important. Thus, a 'nice' and collaborative atmosphere might not at all characterize most relations between firms in a spatial agglomeration. Firms may dislike each other and refuse to talk but can still, indirectly, contribute to each other's competitive success in the global market.
>
> (Malmberg and Maskell 2002: 433)

If this is the case, there are reasons to be cautious about some of the policy initiatives which are inspired by theories of clustering as enhancing collaboration. Further, recent insights on post-human cities (see Chapter 4) also suggest that the vitality of creative areas is an emergent property, created through the correspondence of people, materials and objects, and involving both tacit and explicit knowledges. A wealth of literature should now lead us to the conclusion that the buzz of cities is more than the sum of social interactions between its inhabitants, and that it has more-than-human dimensions which need to be accounted for. The idea that clusters are about the *intense expressivity* of city life is thus important, and one that suggests urban geography can make crucial interventions in economic geography debates.

Irrespective of this, many clusters of creativity (including the music scene in Stockholm) have been considered as identifiable clusters with relatively defined borders. Within these boundaries, dense networks of connection may be built up between firms and individuals. Nonetheless the borders are extremely fluid and the companies involved in each 'cluster' regularly interact and do business with a wide variety of other sectors, types of firms, and customers and collaborators that are not located in the same city (or nation). This hints at another important way of thinking about creative cities: namely the idea they are 'stretched out' places.

FLIGHTPATHS TO EACH OTHER: LOCATING IN CIRCUITS OF CREATIVITY

In debates around knowledge production, there is a significant bifurcation between those who emphasise the importance of face-to-face (and hence, stress the importance of being in specific locales of innovation) and those who feel that the importance of spatial proximity is much overplayed. In relation to the former, many suggest that the type of strong social ties which seem to be important in innovation and creativity can be produced only through spatial proximity which creates opportunities of meetings and mismeetings. Against this, some argue that global communications and air travel allow for a form of stretched-out sociality which far exceeds the limits of a given locale. The suggestion here is that 'being there' is no longer essential to the cultivation of trust or the production of knowledge: in particular, it is argued that virtualism allows for an intimacy and immediacy of communication that previously was possible only face-to-face.

Many writers on learning in the knowledge economy would none-theless contend that the type of social ties maintained at a distance are inevitably 'weak' in form, and that infrequent face-to-face contact reduces the potential for trust and innovation to emerge. The suggestion is that creativity is cultivated through dense local networks. Nonetheless, even those subscribing to the view that face-to-face contact is crucial in fostering innovation concede that global networks are increasingly important in the knowledge economy, with the day-to-day transmission of knowledge (in the form of emails, attachments, faxes, reports) central to the articulation of cultural production. This is taken as evidence of the existence of well-established global 'pipelines' which link firms and individuals, and facilitate the seamless circulation of knowledge across international boundaries.

Accordingly, it appears significant that many cultural clusters emerge in world cities which are defined not by their internal characteristics but by their strategic location in the 'intensified circuits' of the global space economy (see Chapter 5). For instance, when exploring the importance of London as a major European media centre, Gornostaeva and Cheshire (2003) argue that the clustering of creative firms in London is something that can be explained only with reference to the advantageous position of the city within global flows of transport, communication, information and capital. Their suggestion is that localised clusters emerge because of the importance of word-of-mouth in cultural production, but that these clusters are located in places which are well located within more extensive networks of production and transmission.

As such, the co-location of film and television post-production facili-ties in Soho (where 70 per cent of all London video and media firms are situated) is not random, nor is it simply a function of the 'free flow of ideas and learning occuring within the area's many bars, restaurants and coffee-shops' (Nachum and Keeble 2003: 462). Rather, companies pay a premium to locate in Soho because it is a key strategic site of global knowledge production, and not just a cluster serving a city-wide market. Indeed, Nachum and Keeble (2003) show that media firms in Soho may be strongly embedded in the local cluster and rely heavily on resources and processes available locally (not least the skilled and experienced workforce as well as suppliers and customers), yet they maintain linkages extending well beyond the local cluster (especially with the film studios north of London and TV broadcasters in White City and Victoria). Yet Soho is not just conveniently situated in relation to the associated clusters of media production in London, being tied into a network of worldwide media centres, especially Hollywood. In this respect, it is notable that Westminster City Council has recently created a 'wireless zone' in Soho which provides a pervasive high-bandwidth wi-fi network. In turn, this builds upon the transatlantic fibre connections established in the 1990s by Sohonet to link together the advertising, graphic design and post-production companies which constantly need to transmit and circulate design ideas to one another – and around the world (see S. Graham 1999).

Similar global connectivity appears significant in explaining the agglomeration of advertising and creative industries in Manhattan, particularly in the SoHo and Tribeca neighbourhoods which have emerged as dominant global providers of Internet and multimedia skills, design and post-production services. As in London's Soho, these neighbourhoods are gentrifying as they become home to a range of interlocking micro, small and medium-sized digital design, advertising, publishing, fashion, music, multimedia, computing and communications firms. By 1998, there were over 2,200 firms providing more than 50,000 jobs in these sectors (Neff 2005). Here, as with the other advanced producer services (Beaverstock et al. 2000a), the need for face-to-face contact to sustain continuous innovation has been combined with exceptionally high use of advanced telecommunications to link relationally and continuously with the rest of the planet. As a result, Manhattan now has twice the 'domain density' (i.e. concentration of Internet hosts) of the next most 'Internet-rich' city in the United States (Townsend 2001).

Information and communications technology is of course vital to creative firms in an era when media corporations have transnational reach and influence (Robins 1995). Through a process of mergers and acquisitions, these huge media groups have an increasingly global network of branch offices and subsidiaries. This global network of firms within a media group has its *local anchoring points* in those cities that function as 'cultural metropolises' (i.e. as centres of cultural production). While the processes drawing media corporations to world cities are similar to those attracting financial services, it is important to note that leading media cities are not always centres for the providers of global corporate services (Krätke 2002) (see Chapter 5 on the measurement of 'world cityness').

Today's cultural economy is accordingly characterised by not only the globalisation of media corporations but also the spatial concentration of cultural production in a series of metropolitan and world cities (Robins 1995). Cultural globalisation – interpreted as meaning the global proliferation of standardised cultural practices (Appadurai 1996) – cannot proceed without certain *infrastructures* for the dissemination and transmission of cultural forms and commodities on a global scale. As suggested above, these infrastructures comprise telecommunications networks, as well as the organisational and institutional infrastructure of multinational media firms in the culture and media industries. As Krätke (2002) documents, perhaps the most important feature of the global media order is the worldwide network of subsidiaries and branch offices which exists within media corporations and allows them to effectively link urban clusters of cultural production.

The production cluster of the film industry in Babelsberg on the outskirts of Berlin provides further evidence that cluster firms are not only closely networked within the local business area, but also integrated into the supra-regional location networks of *global* media firms. Krätke (2002) demonstrates that in the case of Babelsberg the *local* cluster firms are directly linked with major global media firms from Paris, London and New York and indirectly linked with other global players in the film and TV industry thanks to their proximity to other firms in the area. As Krätke (2002) concludes:

> the global players in the culture industry network locally with the small specialist producers and service providers, while at the same time running a global network of branch offices and subsidiaries that permits global linking of the urban centres of cultural production.
>
> (Krätke 2002: 49)

This networking not only allows global firms to create cultural products which can be tailored to suit local markets, but also facilitates the making of a global community of practice in which ideas and creativity can flourish. One implication here is that creative milieux are creative *because* they constitute a global node in networks of communication and media, with creativity being sparked off by the variety and volume of ideas and images which pass through them.

Despite the emphasis placed in some accounts on the globalised connections, there remains an obdurate myth that while codified knowledge can be dislocated from its originating setting, tacit knowledge is 'context-dependent, spatially-sticky and socially-accessible only through direct physical interaction' (Morgan 2004: 13). Grabher (2002) thus argues that the literature on knowledge production buys into a nested notion of scale whereby local interactions produce tacit knowledges that can be applied to specific projects, but where global pipelines are useful only for transmitting ubiquitous, standardised knowledges. Following Grabher (2002), it seems important to question this distinction, and rethink the role of cities as both *places* and *spaces* of interaction. Here, the work of Amin and Cohendet (2003) spells out an attractive relational way of thinking through the spatiality of urban creativity:

> We question a conceptualisation of knowledge space based on the distinction between place defined as the realm of near, intimate and bounded relations, and space defined as the realm of far, impersonal and fluid relations. It is just this kind of dualism that has allowed commentators to associate tacit knowledge with spatial proximity, and codified knowledge with ubiquity. Against a territorial reading that fashions boundaries around locations on a map, we develop a topological map arranged around relational knowledge associations . . . with individual sites acting as sites of trans-scalar and non-linear connections and as a relay point of circulating knowledges that cannot be territorially attributed with any measure of certainty or fixity.
>
> (Amin and Cohendet 2003: 135)

Here, Amin and Cohendet's frustration with the spatial atomism of the mainstream literature on geographies of knowledge is writ large, and their proposed solution consistent with Richard Smith's (2003) topological take on city (see Chapter 5). Significantly, their summation – the practices of knowing in any single site can no longer be described in terms of a 'local' versus 'global' distinction – draws on the STS literature (see Chapter 4). In effect, they unpack the 'black box' of knowledge production to suggest that the local, or bounded, nature of knowledge actually represents an entanglement of human and non-human practices which are not spatially constrained. This posits creative clusters as 'recipients, combiners, and transmitters' of knowledge which has a variety of sources and takes many forms (Amin and Cohendet 2003).

The implication here is that while there are undeniably localised networks of creativity (e.g. a community of practice made up of employees in a given workplace), creativity typically relies on a more diffuse network of connections:

> The translation of ideas and practices, as opposed to their transmission, are likely to involve people moving to and through 'local' contexts, to which they bring their own blend of tacit and codified knowledges, ways of doing and ways of judging things. There is no one spatial template through which associational understanding or active comprehension takes place. Rather, knowledge translation involves mobile, distanciated forms of information as much as it does proximate relationships.
>
> (Allen 2000b: 28)

Thrift (2000: 685) similarly insists that 'new means of producing creativity and innovation are bound up with new geographies of circulation which are intended to produce situations in which creativity and innovation, can, quite literally, take place'. These 'circulations' include the (now commonplace) global movements of creative workers to attend meetings, seminars, conferences and workshops where exchanging ideas, gossip and knowledge is the order of the day. Such meetings may be fleeting, and on 'neutral' ground, yet allow for the manufacturing of trust and intimacy between individuals who otherwise remain spatially distanced from one another. Bathelt and Schuldt (2005) thus identify international trade fairs and world expos as important opportunities for participants to exchange information about market developments, present new products and monitor innovation. These may also be important for establishing new pipelines which serve to link individuals and firms (see also Bathelt et al. 2004). Given this, it is hardly surprising that city governors are falling over one another in the rush to construct convention centres and conference suites – not just because these are money-spinners (they are often not), but also because it seems to be a prerequisite for connecting a city to global networks of creativity. In this light, city of culture celebrations, festivals and expos can be viewed as attempts to create and sustain creative clusters in a city by establishing global connections (Gold and Gold 2005).

While urban institutions and governors can seek to manufacture these opportunities for global mixing and interaction (and the creation of 'global buzz'), it could of course be argued that this type of mixing goes on in cities anyway. Transnational ties and constant waves of immigration ensure cities are spaces where people from spatially distanced origins meet one another and learn about different cultures. This not only unsettles established knowledges, but also produces new ways of living in and knowing the city:

> City life constantly attracts newcomers, and the trade-mark of newcomers is bringing new ways of looking at things, and maybe new ways of solving old problems. Newcomers are strangers to the city, and things that the old, well-settled residents stopped noticing because of their familiarity, seem bizarre and call for explanation when seen through the eye of a stranger. For strangers, and particularly for the newcomers among them, nothing in the city is 'natural'; nothing is taken for granted by them. Newcomers are born and sworn enemies of tranquillity and self-congratulation. This is not perhaps a situation to be enjoyed by the city natives – but this is also their good luck.
>
> (Bauman 2003: 3)

By definition, creativity involves looking at old problems in new ways, and different ways of seeing may be valuable in this respect. Confirming this, Coe and Bunnell (2003) point to the role of skilled immigrants in the economic development of innovative regions like Silicon Valley (where one-third of the workforce is foreign born). The success of such clusters is predicated on flows of highly skilled migrants, not only because of a shortage of skills locally, but also because they embody different skills sets. Ideas circulate within ethnic, familial and personal networks as much as they do within firms.

Coe and Bunnell (2003) consequently argue that innovation is associated with communities of practice that are best referrred to as 'constellations' which draw together actants operating at different scales. Their view is that the literature on the formation of communities of creative practice can be fruitfully bought into dialogue with the literatures on transnationalism. Within this literature, there has been considerable attention devoted to the heightened salience of migration in the contemporary world, with a particular focus on 'transnational communities'. Deemed as both 'the products of, and catalyst for, contemporary globalisation processes' (Transnational Communities Programme 1998), these communities sit astride political boundaries and, in a very real sense, are neither here nor there. Hannerz (1993) describes four groups of people who fall within this definition. These are, first, 'transnational business' people, including the high-waged, highly skilled professional, managerial and entrepreneurial elites usually associated with finance, banking and business services; second, 'Third World populations', comprising low-waged immigrants who occupy insecure niches in the unskilled or semi-skilled sectors of the urban service economy; third, 'tourists', whose transnational status is often ephemeral but who make up a major proportion of those who are living outside their 'home' space and, finally, 'expressive specialists', in areas such as art, fashion, design, photography, film-making, writing, music and cuisine (A. Scott 2000).

To a lesser or greater extent, all these groups maintain links across long distance and may communicate regularly with family and friends living in a place they once called home. However, Beaverstock and Boardwell (2000) suggest that the designation of some of these groups as living in a state of permanent temporariness may be misleading as many of these groups *settle* in particular cities, often forming 'personal micro-networks' that centre on the expatriate or diasporic residential and leisure-orientated spaces characteristic of major world cities. *Spaces of diaspora* (Box 6.3) thus provide contact zones where different ways of thinking may combine to produce new music, art, food, culture and commerce. In fact, many city authorities actively encourage creativity within such spaces, with a particular focus on the work of 'ethnic entrepreneurs' (Keith 2005).

Box 6.3 SPACES OF DIASPORA

Originally referring to the enforced migration of Jews, the concept of diaspora is currently much in vogue among social scientists. In the process, however, it has become conflated with a number of other categorisations such that it is now applied in a somewhat arbitrary manner to describe all number of immigrant, refugee, guest-worker and traveller identities. In fact, the term diaspora is

continued

commonly used to describe any ethnicised community visible in the host community. Given this definition, it is suggested that it might even apply to gay and lesbian groups, and that it is valid to talk of a queer diaspora, given that sexual dissidents have often been forced to move away from family, home or homeland (Binnie 2004). However, the fact that queer identity has no homeland to identify with problematises this, and suggests that the elasticity of the concept has limits. What is perhaps crucial about diasporic communities is that they share a collective memory and myth about the homeland, including its location, history and achievements, using this to provide a sense of fixity or mooring when living in an unfamiliar environment. This is evident in the social networks and identities created in new contexts away from homelands, as well as the spaces that celebrate diasporic cultures. These cultures may cut across existing boundaries, and challenge notions of authenticity or rootedness. Likewise, these spaces might express a social nearness and closeness to others living thousands of miles away (and concomitant distancing from neighbouring communities and neighbourhoods) (Metcalf 1996). An example here is 'Banglatown', the area of East London that links British Bengali populations with an Asian diaspora (Plate 6.3). Although much of the architecture, street life and activity around Banglatown might be regarded by white British populations as Asian (not least the area's many curry-houses), this is actually a space where notions of British and Bengali identity entwine: in many respects, it constitutes a *heterotopia* whose liminality may spark off new ideas and new ways of being. As a consequence, it is also a space vigorously marketed to tourists: ironically, processes of commodification threaten to erase what is truly distinctive about this space by effacing its diasporic roots (and *routes*).

Further reading: Keith (2005)

Plate 6.3 Brick Lane, Spitalfields – branded as part of Banglatown (photo: Robert House)

Recognising that cities are open to the world and are not (and have never been) 'islands of competitiveness' (Amin 2002: 395) forces us to develop a progressive understanding of the creativity of cities. Herein, it is the sheer 'mixity' of cities which is crucial, with the collision of peoples in the city's multiple public realms bringing new practices into being. This mixity is taken to be a key characteristic of cities in the modern, global era, producing new 'mongrel' forms of urbanism (Sandercock 2003). Yet many urban commentators suggest this mixity is under threat as the city becomes subject to 'logics' of consumerism and privatism. The emergence of 'interdictory spaces' (Flusty 2002) which carve the city into discrete enclaves seems to be eviscerating the mixity of public space, and destroying a mode of metropolitan street life which was unpredictable, sometimes dangerous and dirty, but undeniably fostered creativity. In an era where an ecology of fear is hypothesised to have taken root in the urban West (M. Davis 1998), the question needs to be posed: can the city continue to accommodate difference *and* foster new forms of cultural production?

Here, the current debate around questions of urban conviviality is suggestive of some of the issues at stake. Conviviality is a term which stresses the importance of cultural respect and dialogue, and it contrasts with the agenda of multiculturalism which emphasises respect for cultural difference without resolving the problem of communication between cultures. It may also be contrasted with those versions of cosmopolitanism which suggest cultural difference will ultimately disappear through inter-ethnic mixture and cultural hybridisation (Keith 2005). In the United Kingdom, New Labour favour the former, and have sought to criminalise expressions of religious and ethnic hatred without addressing the question as to how such hatred might be dissipated. In France, debates over the hijab (Muslim headscarf) illustrate another approach, where the state is seeking to efface cultural difference in the name of equality among its citizenry. Though he does not make reference to the latter, it provides a good example of the antagonism that Thrift (2005a) suggests results from different populations being asked to subscribe to notions of citizenship, community or shared order which they regard as fundamentally flawed. Rather than seeking to challenge these political visions by building alternative categories of inclusion and exclusion that make rights-based claims on the state, Thrift (2005a) argues that a more credible and effective form of politics is simply to encourage differently situated individuals and groups to get along. This might require serious thought about the forms of social exchange vital to sustaining the public realm by considering the urban 'props' which give people a reason to interact. It might also require thinking about the types of social networks characteristic of the contemporary era, and recognising that social possibilities of the city's spaces need to be rethought and remediated so that new generations inhabit the city differently (Means and Tims 2005). Conviviality, therefore, is about a living together of different cultures: something that may require new urban spaces where dialogue between strangers is the norm, and not the exception. Whether – and how – these spaces emerge remains to be seen, but without such sites of mixity it seems the creative edge of cities will be blunted – perhaps irrevocably.

CONCLUSION

Contemporary businesses have long subscribed to the idea they can be competitive and creative in the contemporary world if they reject bureaucratic or 'command and control' organisational structures in favour of more pliable, flexible and open forms. On the other hand, urban geographers have been slow to jettison their established ways of thinking about cities, and embrace new ways of approaching questions of what cityness is and what cities do. To these ends, this chapter has considered the debates around urban creativity to demonstrate how new and recent understandings of the city might illuminate a particular urban phenomenon. In some ways, the selection of creativity as a topic of discussion was arbitrary, and less positively valorised issues (e.g. urban crime, poverty, pollution, racism) might have been chosen. Yet the fact that creativity constitutes a buzz concept in current urban studies made it an appropriate topic to explore. This is particularly the case given that certain cities are deemed to have experienced an 'urban renaissance' by virtue of their role as bases for new reflexive forms of consumption and cultural production.

While there are good reasons for doubting the sustainability of the creative industries as engines of economic growth, the literature on urban creativity nonetheless provides many clues as to the distinctive nature of cities as material spaces of production and consumption. As we have seen, if we understand creativity as a routine form of human activity enabled and sustained in specific urban locales, we begin to gain purchase on the profligate materialities of the city. Adapting an ontology of networks, we can hypothesise that the creativity of cities is not produced at any particular scale, but involves different people, knowledges, technologies, discourses, ideas and images which combine in particular urban formations. While this chapter has focused mainly on the creativity found in world cities, this worlding of creativity applies to small towns like Loughborough or Leominster as much as it does to Los Angeles. Creativity is not exclusive to large cities. On the other hand, while it is not exclusive to cities in general, this chapter has suggested that the dense, mixed and stretched materialities of the city are particularly conducive to creativity. Likewise, it has been shown that artistic and cultural activity is endemic to the city, with creativity having made the city at the same time that cities produce culture.

In making this argument, this chapter has sought to underline the argument raised in the opening chapter: namely, that traditional theories of the city, rooted in modern and postmodern thought, pay scant attention to the materiality of cities, and thus neglect much of what is truly distinctive and important about cities. If urban theory is to progress, and to find fitting ways of describing post-millennial cities, a materialist ontology seems an effective way of refocusing our attention on what cities really consist of. Without such a consideration, it is doubtful whether we can answer any of the key ethical and political questions surrounding cities in the twenty-first century.

FURTHER READING

Peter Hall's (1998) *Cities in Civilization* is a massive and wide-ranging account of different moments and spaces of innovation, and while he struggles to isolate what it is about particular cities which causes them to be creative, it remains essential reading. There is a limited literature on art and cities which takes the city seriously, but the case study of London art can be usefully read alongside Jed Perl's (2004) *New Art City*, a detailed account of the rise (and fall) of Manhattan as a critical nexus in international art markets. Overall, the literature on the cultural economy is vast, and within geography, it is the work of Michael Storper and Allen Scott on urban regions that is perhaps best known. Here, Storper (1997) and Scott (2000) provide effective summaries of their respective ideas on the agglomeration of creative know-how in specific neo-Marshallian nodes. There are also many policy-oriented studies of urban creativity which provide a useful overview of the geographies of the culture and media industries: Charles Landry's (2000) *The Creative City* is much cited by policy-makers in the United Kingdom, suggesting that ideas rooted in urban studies often inform policy and practice, while Florida's (2000b) work clearly enjoys global exposure. Finally, it is worth noting that the debates surrounding urban creativity are strangely quiet on questions of gender, with Blake and Hanson (2005) exposing the masculinist assumptions written into much of the literature. It is well worth reading this chapter alongside their article, and thinking about the assumptions commonly made about what creativity is and who is creative.

CONCLUSION

Finding the right vocabulary to describe what is happening in cities is important. Many of the most astute writers on planning and urban geography have talked about how all the great steps forward in conceptual understandings of cities, regions and urban relationships have been based around developing a new vocabulary. Static descriptions of space, for example – empty, crowded, attractive, blighted, residential, light industrial – need to be replaced by more active notions of city geographies and economies, with currents, flows, rhythms, exchanges, transactions and forms of connectedness.

(Worpole 2001: 5)

A sophisticated understanding of the contemporary metropolis . . . demands both empirical awareness of its novelties and a self-consciousness about the historical traces and theoretical burdens that infest the vocabulary of urbanisms through which consideration of the city is rationalised.

(Keith 2005: 29)

This book has reviewed different ways that geographers have made sense of the city, and, in turn, explored how these rest on different theoretical claims about what the city is and what it does. The key argument here has been that urban theory is essentially about forging an appropriate language and vocabulary for describing cities, noting that this terminology informs the way we read, write and practice cities. My key contention is that established languages – particularly those associated with modern and postmodern political economy – are tired, worn-out and, well, boring. This is not to say (as Keith's extract, above, suggests) that we should simply replace the established lexicon of urban studies with an array of neologisms. Rather, it is to insist that geographers question established concepts and jettison those which are no longer useful for making sense of the urban experience. Being reflexive about how we study what we study is thus vital. So too is an awareness that our understanding of cities can only ever be partial: though some might try, there can never be a unified theory of cities, only knowledges written from particular viewing angles in particular places and times.

That said, in Chapters 2–6 this book has dwelt on the *materiality* of cities, suggesting that a rejuvenated urban theory might emerge from a more serious consideration of the city's diverse materiality. Problematising extant distinctions of image/reality, discourse/practice, nature/culture and local/global, it has been contended that a focus on materiality grants us purchase on what is truly urban about city life, and stakes out an agenda for urban studies which takes the city seriously as an object of study. The focus of this book has been on the contemporary West, yet many (if not all) of the ideas in the latter chapters of this book are relevant to the study of non-Western cities too. After all, suggesting the city is comprised of a more-or-less dense combination of 'stuff' – technologies, texts, people, animals, computers, plants, words and images – and that this stuff exists in more-or-less expansive networks of connection, provides an ontological framework for thinking about cityness which does not privilege any particular notion of structure or agency. Nor does it assume that people are makers of cities any more than animals, or software, or pathogens. But what it does is to impel us to explore the way these materials combine in particular instances with particular forces, and to scrutinise how this play of effects and affects produces particular urban formations. As always, there remains much urban geography to be done.

BIBLIOGRAPHY

Adams, J. (1999) *The Social Implications of Hypermobility*. Working paper delivered to Economic and Social Consequences of Sustainable Pollution conference. Paris: OECD.

ADERLY, (l'Agence pour le Développement Economique de la Région Lyonnaise) (1997) *Lyon et sa Region: Faits et Chiffres*. Lyon: ADERLY.

Adey, P. (2004) Surveillance at the airport, *Environment and Planning A* 36: 1365–1380.

Aitken, S. (2005) Textual analysis, in Flowerdew, R. and Martin, D. (eds) *Methods in Human Geography*. Harlow: Pearson.

Albrechts, G. and Lievois, G. (2004) The Flemish diamond: urban network in the making? *European Planning Studies* 12(3): 350–372.

Allen, J. (1995) Crossing borders: footloose multinationals?, in Allen, J. and Hamnett, C. (eds) *A Shrinking World?* Oxford: Oxford University Press.

Allen, J. (2000a) On Georg Simmel: proximity, distance and movement, in Crang, M. and Thrift, N. (eds) *Thinking Space*. London: Routledge.

Allen, J. (2000b) Power/economic knowledges: symbolic and spatial formations, in Bryson, J., Daniels, P.W., Henry, N. and Pollard, J. (eds) *Knowledge, Space, Economy*. London: Routledge.

Allen, J., Massey, D. and Pryke, M. (2001) *Unsettling Cities: Movement/Settlement*. London: Routledge.

Alonso, W. (1964) *Location and Land Use*. Cambridge, MA: Harvard University Press.

Amin, A. (2002) Spatialities of globalisation, *Environment and Planning A* 34(3): 385–399.

Amin, A. (2004) Regions unbound: towards a new politics and place, *Geografiska Annaler B* 86(1): 31–42.

Amin, A. and Cohendet, P. (2003) *Knowledge Practices: Communities and Competences in Firms*. Oxford: Oxford University Press.

Amin, A. and Graham, S. (1998) The ordinary city, *Transactions of the Institute of British Geographers*, NS22(4): 411–429.

Amin, A. and Thrift, N. (1992) Neo-Marshallian nodes in global networks, *International Journal of Urban and Regional Research* 16(4): 571–587.

Amin, A. and Thrift, N. (2002) *Cities: Reimagining the Urban*. Cambridge: Polity.

Amin, A., Massey, D. and Thrift, N. (2000) *Cities for the Many, Not the Few*. Bristol: Policy Press.

Anderson, B. (2004) Recorded music and the practices of remembering, *Journal of Social and Cultural Geography* 5(1): 3–20.

Anderson, B. and Tolia-Kelly, D. (2004) Matter(s) in social and cultural geography, *Geoforum* 35(5): 669–674.

Anderson, K. (1991) *Vancouver's Chinatown: Racial Discourse in Canada, 1875–1980*. Montreal: McGill-Queen's University Press.

Anderson, K. (1995) Culture and nature at the Adelaide Zoo: at the frontiers of 'human' geography, *Transactions, Institute of British Geographers* 20: 275–294.

Anderson, K. (2000) 'The beast within': race, humanity and animality, *Environment and Planning D: Society and Space* 18: 301–320.

Anderson, K. and Gale, F. (eds) (1992) *Inventing Places: Studies in Cultural Geography*. Melbourne: Longman Cheshire.

Appadurai, A. (1996) *Modernity at Large: Cultural Dimensions of Globalization*. Minneapolis, MN: University of Minnesota Press.

Atkinson, D. and Cosgrove, D.E. (1998) Urban rhetoric and embodied identities: city, nation, and empire at the Vittorio Emanuele II Monument in Rome, 1870–1945, *Annals of the Association of American Geographers* 88(1): 28–49.

Augé, M. (1995) *Non-places: Introduction to an Anthropology of Supermodernity*. London: Verso.

Baigent, E. (2004) Patrick Geddes, Lewis Mumford and Jean Gottmann: divisions over 'megalopolis', *Progress in Human Geography* 28(6): 687–700.

Bailey, N. and Turok, I. (2001) Central Scotland as a polycentric urban region: useful planning concept or chimera?, *Urban Studies* 38(7): 697–715.

Barber, S. (2002) *Projected Cities*. London: Reaktion.

Barnes, T. (2001) Retheorising economic geography: from the quantitative revolution to the cultural turn, *Annals, Association of American Geographers* 92: 546–565.

Barnes, T. (2005) Culture: economy, in Cloke, P. (ed.) *Spaces of Geographical Thought*. London: Sage.

Barnes, T. and Duncan, J. (eds) (1992) *Writing Worlds*. London: Routledge.

Barnes, T. and Gregory, D. (1997) *Reading Human Geography: The Poetics and Politics of Inquiry*. London: Arnold.

Barthes, R. (1975) *Mythologies*. Paris: Seuil.

Bassett, K., Griffiths, R. and Smith, I. (2003) City of culture?, in Boddy, M. (ed.) *Urban Transformation and Urban Governance: Shaping the Competitive City of the Future*. Bristol: Policy Press.

Bathelt, H. and Schuldt, N. (2005) Between luminaries and meat-grinders: international trade fairs as temporary clusters, paper presented to Association of American Geographers Annual Meeting, Denver, CO.

Bathelt, H., Mamberg, A. and Maskell, P. (2004) Clusters and knowledge: local buzz, global pipelines and the process of knowledge creation, *Progress in Human Geography* 28(1): 31–56.

Baudrillard, J. (1990) *Cool Memories*. London: Verso.

Bauman, Z. (2000) *Liquid Modernity*. Cambridge: Polity.

Bauman, Z. (2001) Consuming life, *Journal of Consumer Studies* 1(1): 9–29.

Bauman, Z. (2003) *City of Fears, City of Hopes*, working paper. London: Goldsmiths College, University of London.

Beaverstock, J.V. and Boardwell, J.T. (2000) Negotiating globalization, transnational corporations and global city financial centres in transient migration studies, *Applied Geography* 20(2): 227–304.

Beaverstock, J.V., Smith, R.G., Taylor, P.J., Walker, D.R.F. and Lorimer, H. (2000a) Globalization and world cities: some measurement methodologies, *Applied Geography* 20(1): 43–63.

Beaverstock, J.V., Taylor, P.J. and Smith, R.G. (2000b) The long arm of the law: London's law firms in a globalising world economy, *Environment and Planning A* 31(10): 1857–1876.

Bech, H. (1998) Citysex: representing lust in public, *Theory, Culture and Society* 15: 215–241.

Bell, D. and Binnie, J. (2004) Authenticating queer space: citizenship, urbanism and governance, *Urban Studies* 41: 1807–1820.

Bender, B. (ed.) (1992) *Landscape: Politics and Perspectives*. Oxford: Berg.

Bendle, M. (2002) The crisis of 'identity' in high modernity, *British Journal of Sociology* 53(1): 1–18.

Benjamin, W. (1985) *One Way Street and Other Writings*. London: Verso.

Benjamin, W. (1991) *Charles Baudelaire: A Lyric Poet in the Era of High Capitalism*. London: Verso.

Benjamin, W. (1999) *The Arcades Project*. Cambridge, MA: Belknap Press.

Berger, J. (1977) *Ways of Seeing*. London: Penguin.

Berman, M. (1983) *All That is Solid Melts into Air: The Experience of Modernity*. New York: Verso.

Berry, B.J.L. (1964) Cities as systems of cities within systems of cities, *Papers of the Regional Science Association* 13: 147–163.

Berry, B.J.L. (1967) *Geography of Market Centres and Retail Distribution*. Englewood Cliffs, NJ: Prentice-Hall.

Berry, B.J.L. and Rees, P.H. (1969) Factorial ecology of Calcutta, *American Journal of Sociology* 74: 445–491.

Bingham, N. and Thrift, N. (2000) Michael Serres and Bruno Latour, in Crang, M. and Thrift, N. (eds) *Thinking Space*. London: Routledge.

Binnie, J. (2004) *The Globalization of Sexuality*. London: Sage.

Binnie, J. and Skeggs, B. (2004) Cosmopolitan knowledge and consumption of sexualised space: Manchester's gay village, *Sociological Review* 52(1): 39–61.

Binnie, J. and Valentine, G. (1999) Geographies of sexuality: a review of progress, *Progress in Human Geography* 23: 175–187.

Black, I.S. (1999) Rebuilding *The Heart of the Empire*: bank headquarters in the City of London, 1919–1939, reprint in Arnold, D. (ed.) *The Metropolis and its Image: Constructing Identities for London, c. 1750–1950*. Oxford: Blackwell.

Blake, M. and Hanson, S. (2005) Rethinking innovation: context and gender, *Environment and Planning A* 37(5): 681–701.

Blunt, A. and Wills, J. (2000) *Dissident Geographies: An Introduction to Radical Ideas and Practice*. Harlow: Prentice Hall.

Boddy, M. (1999) Geographical economics and urban competitiveness, *Urban Studies* 36(5–6): 811–842.

Bogard, W. (2002) Simmel in cyberspace: strangeness and distance in postmodern communications, *Space and Culture* 4–5: 23–46.

Bondi, L. (1991) Gender divisions and gentrification: a critique, *Transactions of the Institute of British Geographers* 16(1): 157–170.

Bondi, L. and Christie,H. (2002) Gender relations and the city, in Bridge, G. and Watson, S. (eds) *The City Companion*. Oxford: Blackwell.

Booth, C. (1889) *Life and Labour of the People in London*, vol. 1. London: Macmillan.

Borden, I. (2001) *Skateboarding, Space and the City: Architecture and the Body*. Oxford: Berg.

Boyle, M. and Hughes, C.G. (1995) The politics of urban entrepreneurialism in Glasgow, *Geoforum* 25(4): 453–470.

Brenner, N. (1998) Between fixity and motion: accumulation, territorial organization and the historical geography of spatial scales, *Environment and Planning D: Society and Space* 16(5): 459–481.

Brenner, N. (1999) Globalisation as reterritorialisation: the re-scaling of urban governance in the European Union, *Urban Studies* 36(3): 431–451.

Brosseau, M. (1994) Geography's literature, *Progress in Human Geography* 18: 333–353.

Brown, M. (2001) *Closet Space*. London: Routledge.

Browne, K. (2004) Genderism and the bathroom problem: (re)materializing sexed sites, (re)creating sexed bodies, *Gender, Place and Culture* 11(3): 331–352.

Burgess, J.A. (1985) News from nowhere: the press, the riots and the myth of the inner city, in Burgess, J.A. and Gold, J. (eds) *Geography, the Media and Popular Culture*. London: Croom Helm.

Burgess, J.A. and Gold, J. (eds) (1985) *Geography, the Media and Popular Culture*. London: Croom Helm.

Butler, R. (1999) The body, in Cloke, P., Crang, M. and Goodwin, M. (eds) *Introducing Human Geographies*. London: Arnold.

Butler, T. (2003) Gentrification and its others, *Urban Studies* 40(11): 2469–2486.

Callard, F. (1998) Affective histories: the shape of anxiety in London 1880–1914, paper presented to IBG-RGS conference, Kingston, Surrey.

Callon, M., Lascoumes, P. and Barthe, Y. (2001) *Agir dans un monde incertain: essai sur la democratie technique*. Paris: Seuil.

Carter, H. (1988) *An Introduction to Urban Historical Geography*. London: Arnold.

Castells, M. (1977) *The Urban Question*. London: Arnold.

Castells, M. (1978) *City, Class and Power*. London: Arnold.

Castells, M. (1983) *The City and the Grassroots: A Cross-cultural Theory of Urban Social Movements*. London: Arnold.

Castells, M. (1989) *The Informational City: Information Technology, Economic Restructuring and the Urban-Regional Process*. Oxford: Blackwell.

Castells, M. (1996) *The Information Age: Economy, Society, and Culture, Volume 1: The Rise of the Network Society*. Oxford: Blackwell.

Castells, M. (1997) *The Information Age: Economy, Society, and Culture, Volume 2: The Power of Identity*. Oxford: Blackwell.

Castells, M. (1998a) *The End of the Millennium*. Oxford: Blackwell.

Castells, M. (1998b) *The Information Age: Economy, Society, and Culture, Volume 3: End of the Millennium*. Oxford: Blackwell.

Castells, M. (1999) The informational city is a dual city: can it be reversed?, in Scohn, D.A., Sanyal, B. and Mitchell, W. (eds) *High Technology and Low-income Communities*. Cambridge, MA: MIT Press.

Castells, M. (2000) Materials for an exploratory theory of the network society, *British Journal of Sociology* 51 (1): 1–24.

Castells, M. (2001) *The Internet Galaxy*. Oxford: Oxford University Press.

Castells, M. (2002a) Urban sociology for the twenty-first century, in Susser, I. (ed.) *The Castells Reader on Cities and Social Theory*. Oxford: Blackwell.

Castells, M. (2002b) Local and global: cities in the network society, *Tijdschrift voor Economische en Sociale Geografie* 93 (5): 548–558.

Castells, M. and Hall, P. (1994) *Technopoles of the World: The Making of 21st Century Industrial Complexes*. London: Routledge.

Castree, N. (1999) Envisioning capitalism: geography and the renewal of Marxian political economy, *Transactions of the Institute of British Geographers* 24 (2): 324–340.

Castree, N. (2005) *Nature*. London: Routledge.

Chant, C. and Goodman, D. (eds) (1998) *Pre-industrial Cities and Technology*. London: Routledge.

Chant, C. and Goodman, D. (eds) (2000) *European Cities and Technology Reader: Industrial to Post-industrial City*. London: Routledge.

Cherry, G. (1988) *Cities and Plans*. London: Arnold.

Cheshire, P. (1999) Cities in competition: articulating the gains from integration, *Urban Studies* 36 (5–6): 843–864.

Childe, G. (1936) *Man Makes Himself*. London: Watts.

Church, A. and Reid, P. (1996) Urban power, international networks and competition: the example of cross-border co-operation, *Urban Studies* 33 (8): 1297–1318.

Clark, D. (1996) *Urban World, Global City*. London: Routledge.

Clark, W.A.V. (2002) Monocentric to polycentric: new urban forms and old paradigms, in Bridge, G. and Watson, S. (eds) *The City Companion*. Oxford: Blackwell.

Clarke, D.B. (1997) *The Cinematic City*. London: Routledge.

Clarke, D.B. (2003) *The Consumer Society and the Postmodern City*. London: Routledge.

Clarke, P. (1989) The economic currency of architectural aesthetics, in Diani, M. and Ingraham, C. (eds) *Restructuring Architectural Theory*. Evanston, IL: Northwestern University Press.

Clifton, R. and Maughan, E. (2000) *Twenty-Five Visions: The Future of Brands*. London: Macmillan.

Cloke, P. and Jones, O. (2002) *Tree Cultures: The Place of Trees and Trees in their Place*. Oxford: Berg.

Cloke, P. and Jones, O. (2004) Turning in the graveyard: trees and the hybrid geographies of dwelling, monitoring and resistance in a Bristol cemetery, *Cultural Geographies* 11(2): 313–341.

Cloke, P., Philo, C. and Sadler, D. (1991) *Approaching Human Geography*. London: Chapman.

Coaffee, J. and Wood, D. (2006) Security and surveillance, in Hubbard, P., Hall, T. and Short, J.R. (eds) *The Urban Compendium*. London: Sage.

Coe, N. and Bunnell, T. (2003) 'Spatializing' knowledge communities: towards a conceptualization of transnational innovation networks, *Global Networks* 3(4): 437–456.

Coe, N. and Townsend, A.R. (1998) Debunking the myth of localized agglomerations: the development of a regionalized service economy in South-East England, *Transactions of the Institute of British Geographers* 23: 385–404.

Coleman, A. (1985) *Utopia on Trial*. London: Croom Helm.

Cook, M. (2003) *London and the Culture of Homosexuality 1885–1914*. Cambridge: Cambridge University Press.

Cook, S.D.N. and Seely-Brown, K. (1999) Bridging epistemologies: the generative dance between organisational knowledge and organisational knowing, *Organisational Science* 10(4): 381–400.

Cooke, P. (1983) *Theories of Planning and Spatial Development*. London: Hutchinson.

Cooke, P. (1989) *Localities*. London: Routledge.

Corbin, A. (1998) *Village Bells*. London: Pimlico.

Cosgrove, D. (1984) *Social Formation and Symbolic Landscape*. London: Croom Helm.

Cosgrove, D. (1989) Geography is everywhere: culture and symbolism in human landscapes, in Gregory, D. and Walford, R. (eds) *Horizons in Human Geography*. London: Macmillan.

Cosgrove, D. and Daniels, S. (1988) *The Iconography of Landscape*. Cambridge: Cambridge University Press.

Cox, K. (1991) Questions of abstraction in studies in the new urban politics, *Journal of Urban Affairs* 13: 267–280.

Cox, K. (1993) The local and the global in the new urban politics: a critical view, *Environment and Planning D: Society and Space* 11: 433–448.

Cox, K. (1995) Globalisation, competition and the politics of local economic development, *Urban Studies* 32(2): 213–225.

Cox, K. and Mair, A. (1989) Levels of abstraction in locality studies, *Antipode* 21(2): 121–132.

Crang, M. (1998) *Cultural Geography*. London: Routledge.

Crang, M. (2002) Between places: producing hubs, flows and networks, *Environment and Planning A*, 34(4): 569–574.

Cresswell, T. (1996) *In Place/Out of Place: Geography, Ideology and Transgression*. Minneapolis, MN: University of Minnesota.

Cresswell, T. (1999) Night discourse, in Fyfe, N. (ed.) *Images of the Street*. London: Routledge.

Cresswell, T. (2001) The production of mobilities, *New Formations* 43: 11–25.

Cresswell, T. (2003) Landscape and the obliteration of practice, in Anderson, K., Domosh, M., Pile, S. and Thrift, N. (eds) *Handbook of Cultural Geography*. London: Sage.

Crewe, L. and Beaverstock, J. (1998) Fashioning the city: cultures of consumption in contemporary urban spaces, *Geoforum* 29(3): 287–308.

Crewe, L., Gregson, N. and Brooks, K. (2003) Alternative retail spaces, in Leyshon, A. and Lee, R. (eds) *Alternative Economic Geographies*. London: Sage.

Crouch, D. (2001) Spatialities and the feeling of doing, *Social and Cultural Geography* 2(1): 61–75.

Crouch, D. (2003) Performances and constitutions of natures: a consideration of the performance of lay geographies, *The Sociological Review* 51: 17–30.

Daly, G. (1998) Homeless and the street: observations from Britain and the US, in Fyfe, N. (ed.) *Images of the Street: Planning, Identity and Control*. London: Routledge.

Daniels, S. (1993) *Fields of Vision: Landscape Imagery and National Identity in England and the United States*. Princeton, NJ: Princeton University Press.

Davidson, J. (2000) '. . . the world is getting smaller': women, agoraphobia and bodily boundaries, *Area* 32(1): 31–40.

Davidson, J. and Bondi, L. (2004) Spatialising affect; affecting space: introducing emotional geographies, *Gender Place and Culture* 11: 373–374.

Davies, G. (2003) A geography of monsters?, *Geoforum* 34(4): 409–412.

Davis, M. (1990) *City of Quartz: Excavating the Future in Los Angeles*. New York: Verso.

Davis, M. (1998) *The Ecology of Fear: Los Angeles and the Imagination of Disaster*. London: Verso.

Davis, R. (1973) Urban trends, *Atlantic Quarterly* 23(3): 34–38.

Davoudi, S. (2002) Polycentricity: modelling or determining reality, *Town and Country Planning* 71: 114–119.

Dear, M. (2000) *The Postmodern Urban Condition*. Oxford: Blackwell.

Dear, M. (2002) Los Angeles and the Chicago School: invitation to a debate, *City and Community* 1(1): 5–32.

Dear, M. and Flusty, S. (1998) Postmodern urbanism, *Annals of the Association of American Geographers* 88: 50–72.

Deas, I.A. and Giordano, B. (2001) Conceptualising and measuring urban competitiveness: the relationship between assets and outcomes in major English cities, *Environment and Planning A* 33: 1411–1429.

Deas, I.A. and Ward, K.G. (2000) From the 'new localism' to the 'new region-alism'? The implications of regional development agencies for city-regional relations, *Political Geography* 19(3): 273–292.

Debord, G. (1967) *La Societé du spectacle*. Paris: Buchet-Chastel.

de Certeau, M. (1984) *The Practices of Everyday Life*. Berkeley, CA: University of California Press.

de Certeau, M. (2000) Walking in the city, in Ward, G. (ed.) *The de Certeau Reader*. London: Blackwell.

Del Casino, V. and Hanna, M. (2000) Representations and identities in tourist map spaces, *Progress in Human Geography* 20(1): 23–46.

Deleuze, G. and Guattari, F. (1987) *A Thousand Plateaus: Capitalism and Schizophrenia*. Minneapolis, MN: University of Minnesota Press.

Department for Culture, Media and Sport (DCMS) (2001) *The Creative Industries Mapping Document*. London: DCMS.

Derrida, J. (1991) *Writing and Difference*. London: Routledge.

Derudder, B. and Witlox, F. (2005) An appraisal of the use of airline data in assessing the world city network: a research note on data, *Urban Studies* 42(13): 2371–2388.

Derudder, B., Vereecken, L. and Witlox, F. (2004) *An Appraisal of the Use of Airline Data in Assessments of the World City Network*, GaWc bulletin 152.

de Vries, G. (2001) Big city, great art: a myth about art production, in Klamer, A. (ed.) *The Value of Culture*. Amsterdam: Amsterdam University Press.

Dodge, M. and Kitchin, R. (2000) *The Atlas of Cyberspace*. London: Continuum.

Dodge, M. and Kitchin, R. (2005) Code and the transduction of space, *Annals, Association of American Geographers* 95(1): 162–180.

Doel, M.A. (1999) *Poststructuralist Geographies: The Diabolical Art of Spatial Science*. Edinburgh: Edinburgh University Press.

Doel, M.A. (2004) Post-structuralism: the essential selection, in Cloke, P. (ed.) *Envisioning Human Geography*. London: Arnold.

Doel, M. and Clarke, D. (2004) Paul Virilio, in Hubbard, P., Kitchin, R. and Valentine, G. (eds) *Key Thinkers on Space and Place*. London: Sage.

Domosh, M. (1989) A method for interpreting landscape: a case study, *Area* 21(4): 347–355.

Donald, J. (1999) *Imagining the Modern City*. London: Athlone.

Drake, G. (2003) This place gives me space: place and creativity in the creative industry, *Geoforum* 34(4): 511–524.

Driver, F. (1988) Moral geographies, social science and the urban environment in the mid-nineteenth century, *Transactions, Institute of British Geographers* 13(2): 275–287.

Driver, F. and Gilbert, D. (1999) *Imperial Cities: Landscape, Display and Identity*. Manchester: Manchester University Press.

Duncan, J. (1990) *The City as Text: The Politics of Landscape Interpretation in the Kandyan Kingdom*. Cambridge: Cambridge University Press.

Duncan, J. and Ley, D. (eds) (1993) *Place/Culture/Representation*. London: Routledge.

Durkheim, E. (1893) [1997] *The Division of Labour in Society.* New York: Free Press.

Edensor, T. (2000) Walking in the countryside: reflexivity, embodied practices and ways to escape, *Body and Society* 6(3–4): 81–106.

Edensor, T. (2005) Waste matter: the debris of industrial ruins and the disordering of the material world, *Journal of Material Culture* 10(3): 311–332.

Engels, F. (1844) [1887] *The Condition of the Working-Class in England.* London: Swan Sonnenschein.

Farish, M. (2005) Cities in shade: urban geography and the uses of noir, *Environment and Planning D: Society and Space* 23(1): 95–118.

Fishman, R. (1977) *Urban Utopias in the Twentieth Century: Ebenezer Howard, Frank Lloyd Wright, and Le Corbusier.* New York: Basic Books.

Florida, R. (2002a) The economic geography of talent, *Annals, Association of American Geographers* 92(4): 743–755.

Florida, R. (2002b) *The Rise of the Creative Class.* New York: Basic Books.

Florida, R. (2002c) Bohemia and economic geography, *Journal of Economic Geography* 2(1): 255–271.

Flowerdew, J. (2004) The discursive construction of a world class city, *Discourse and Society* 15: 579–604.

Flusty, S. (2000) Thrashing Downtown: play as resistance to the spatial and representational regulation of Los Angeles, *Cities* 17(2): 149–158.

Forel, A. and Fetscher, R. (1931) *The Sexual Question.* Batavia, Germany: Geode Tempelerian.

Foucault, M. (1967) *Madness and Civilisation: A History of Insanity in an Age of Reason.* London: Tavistock.

Foucault, M. (1980) *Power/Knowledge: Selected Interviews and Other Writings 1972/1977,* ed. C. Gordon. Brighton: Harvester.

Freeman, C. and Louca, F. (2002) *As Time Goes By: From the Industrial Revolutions to the Information Revolution.* Oxford: Oxford University Press.

Friedberg, A. (2002) Urban mobility and cinematic visuality: the screens of Los Angeles – endless cinema or private telematics, *Journal of Visual Culture* 1(2): 183–204.

Friedmann, J. (1986) The world city hypothesis, *Development and Change* 17(1): 69–83.

Friedmann, J. (2000) Reading Castells: *Zeitdiagnose* and social theory, *Environment and Planning D: Society and Space* 18(1): 111–120.

Friedmann, J. and Wolff, G. (1982) World city formation: an agenda for research and action, *International Journal of Urban and Regional Research* 3(2): 309–344.

Fyfe, N. (1996) Contested visions of a modern city: planning and poetry in postwar Glasgow, *Environment and Planning A* 28(3): 345–367.

Fyfe, N. (ed.) (1998) *Images of the Streets: Planning, Identity and Control in Public Space.* London: Routledge.

Fyfe, N. (2004) Zero Tolerance, maximum surveillance, in Lees, L. (ed.) *The Emancipatory City: Paradoxes and Possibilities.* London: Sage.

Fyfe, N. and Bannister, J. (1996) City watching: closed circuit television surveillance in public spaces, *Area* 28: 37–46.

Gaffikin, F. and Warf, B. (1993) Urban policy and the post-Keynesian state in the United Kingdom and the United States, *International Journal of Urban and Regional Research* 17(1): 67–84.

Gandy, M. (2004a) Rethinking urban metabolism: water, space and modern city, *City* 8(3): 363–379.

Gandy, M. (2004b) Water, modernity and emancipatory urbanism, in Lees, L. (ed.) *The Emancipatory City: Paradoxes and Possibilities.* London: Sage.

Gandy, M. (2005) Cyborg urbanization: complexity and monstrosity in the contemporary city, *International Journal of Urban and Regional Research* 29(1): 26–49.

Garreau, J. (1991) *The Edge City: Life on the New Frontier.* New York: Doubleday.

Geddes, P. (1915) *Cities in Evolution.* London: Williams & Norgate.

Gelder, T. and Thornton, S. (1997) *The Subcultures Reader.* London: Routledge.

George, H. (1884) *Social Problems.* London: Kegan Paul.

Gibson-Graham, J.K. (1996) *The End of Capitalism as We Knew it: A Feminist Critique of Political Economy.* Oxford: Blackwell.

Giddens, A. (1990) *The Consequences of Modernity.* Cambridge: Polity Press.

Gilbert, D. and Henderson, F. (2002) London and the tourist imagination, in Gilbert, P. (ed.) *Imagined Londons.* Albany, NY: State University of New York Press.

Glaeser, E. (2000) The new economics of urban and regional growth, in Clark, G., Gertler, M. and Feldman, M. (eds) *The Oxford Handbook of Economic Geography.* Oxford: Oxford University Press.

Gleeson, B. (2001) Disability and the open city, *Urban Studies* 38(2): 251–265.

Godfrey, B. (1988) *Neighborhoods in Transition: The Making of San Francisco's Ethnic and Nonconformist Communities.* Berkeley, CA: University of California Press.

Goffman, E. (1959) *The Presentation of Self in Everyday Life.* Harmondsworth: Penguin.

Gold, J.R. (1998) The death of the boulevard, in Fyfe, N. (ed.) *Images of the Street.* London: Routledge.

Gold, J.R. and Gold, M. (2005) *Cities of Culture.* Chichester: Ashgate.

Gold, J.R. and Ward, S.V. (1994) *Place Promotion: The Use of Publicity and Marketing to Sell Towns and Regions.* Chichester: Spon.

Gornostaeva, G. and Cheshire, P. (2003) *Economic Performance in European Regions Media Cluster in London.* Paris: Cahiers de l'Aurif 135.

Goss, J. (1993) The 'magic of the mall': an analysis of form, function, and meaning in the contemporary retail built environment, *Annals of the Association of American Geographers* 83: 18–47.

Gottdiener, M. and Budd, L. (2005) *Key Concepts in Urban Studies.* London: Sage.

Gottdiener, M. and Lagopoulus, A. (1986) *The City and the Sign: An Introduction to Urban Semiotics.* New York: Columbia University Press.

Gottmann, J. (1987) *Megalopolis Revisited: 25 Years Later*. College Park, MD: Institute for Urban Studies, University of Maryland.

Grabher, G. (2001) Ecologies of creativity: the Village, the Group and the heterarchic organisation of the British advertising industry, *Environment and Planning A* 33: 351–374.

Grabher, G. (2002) Cool projects, boring institutions: temporary collaborations in social context, *Regional Studies* 36: 205–214.

Graham, B. (2002) Heritage as knowledge: Capital or culture?, *Urban Studies* 39(5–6): 1003–1017.

Graham, S. (1999) Global grids of glass: on global cities, telecommunications and planetary urban networks, *Urban Studies* 36(5–6): 929–949.

Graham, S. (2004a) Beyond the 'dazzling light': from dreams of transcendence to the 'remediation' of urban life, *New Media and Society* 6(1): 16–25.

Graham, S. (2004b) Excavating the material geographies of cybercities, in Graham, S. (ed.) *The Cybercities Reader*. London: Routledge.

Graham, S. (ed.) (2004c) *The Cybercities Reader*. London: Routledge.

Graham, S. (2005) Software-sorted geographies, *Progress in Human Geography* 29(5): 562–580.

Graham, S. and Marvin, S. (1995) More than ducts and wires: post fordism, cities and utility networks, in Healey, P., Cameron, S., Davoudi, S., Graham, S. and Madani-Pour, A. (eds) *Managing Cities: The New Urban Context*. Chichester: John Wiley.

Graham, S. and Marvin, S. (2001) *Splintering Urbanism*. London: Routledge.

Greed, C. (1994) *Women and Planning: Creating Gendered Realities*. London: Routledge.

Grenfell, M. and Hardy, C. (2003) Field manoeuvres: Bourdieu and the Young British Artists, *Space and Culture* 6: 19–34.

Griffiths, H., Poulter, I. and Sibley, D. (2000) Feral cats in the city, in Philo, C. and Wilbert, C. (eds) *Animal Spaces, Beastly Places: New Geographies of Human–Animal Relations*. London: Routledge.

Groth, J. and Corijn, E. (2005) Reclaiming urbanity, *Urban Studies* 42(3): 503–526.

Habermas, J. (1989) *The Structural Transformation of the Public Sphere: An Inquiry into a Category of Bourgeois Society*. Cambridge, MA: MIT Press.

Hackworth, J. and Smith, N. (2001) The changing state of gentrification, *Tijdschrift voor Economische en Sociale Geografie* 92(4): 464–477.

Halfacree, K.H. and Kitchin, R.M. (1996) 'Madchester rave on': placing the fragments of popular music, *Area* 28(1): 49–55.

Hall, P.S. (1966) *The World Cities*. London: Weidenfeld and Nicolson.

Hall, P.J. (1988) *Cities of Tomorrow: An Intellectual History of Urban Planning and Design in the Twentieth Century*. New York: Basil Blackwell.

Hall, P.J. (1998) *Cities in Civilization*. London: Weidenfeld and Nicolson.

Hall, P.J. (2001) Global city-regions in the twenty-first century, in Scott, A.J. (ed.) *Global City-Regions: Trends, Theory, Policy*. Oxford: Oxford University Press.

Hall, P.J. (2005) How cities became great, *Newsweek* 25 April: 17.

Hall, S. (1980) Encoding/decoding, in Centre for Contemporary Cultural Studies (eds) *Culture, Media, Language: Working Papers in Cultural Studies, 1972–79*. London: Hutchinson.

Hall, T. (1997) Images of industry in the post-industrial city: Raymond Mason and Birmingham, *Ecumene* 4(1): 46–68.

Hall, T. (2001) *Urban Geography*, 2nd edn. London: Routledge.

Hall, T. (2003) Car-ceral cities: social geographies of everyday urban mobility, in Miles, M. and Hall, T. (eds) *Urban Futures*. London: Routledge.

Hall, T. (2004) Public art, civic identity and the new Birmingham, in Kennedy, L. (ed.) *Remaking Birmingham: The Visual Culture of Urban Regeneration*. London: Routledge.

Hall, T. and Hubbard, P. (1996) The entrepreneurial city: new urban politics, new urban geographies?, *Progress in Human Geography* 20(2): 153–174.

Hall, T. and Hubbard, P. (eds) (1998) *The Entrepreneurial City: Geographies of Politics, Regime, and Representation*. Chichester: John Wiley.

Hamnett, C. (1991) The blind men and the elephant: the explanation of gentrification, *Transactions of the Institute of British Geographers* 16: 173–189.

Hamnett, C. (2003) Contemporary human geography: fiddling while Rome burns?, *Geoforum* 34(1): 1–3.

Hannerz, U. (1993) The cultural roles of world cities, in Cohen, A.P. and Fukuo, K. (eds) *Humanizing the City*. Edinburgh: Edinburgh University Press.

Haraway, D. J. (1991) *Simians, Cyborgs and Women: The Reinvention of Nature*. London: Free Association Books.

Harding, A. (1994) Urban regimes and growth machines: towards a cross-national research agenda, *Urban Affairs Quarterly* 29(1): 33–58.

Harley, J.B. (1992) Deconstructing the map, in Barnes, T. and Duncan, J. (eds) *Writing Worlds: Discourse, Text and Metaphor in the Representation of Landscape*. London: Routledge.

Harloe, M. (1977) *Captive Cities*. London: John Wiley.

Harvey, D. (1969) *Explanation in Geography*. London: Arnold.

Harvey, D. (1973) *Social Justice and the City*. London: Arnold.

Harvey, D. (1982) *The Limits to Capital*. Oxford: Blackwell.

Harvey, D. (1985) The geopolitics of capitalism, in Gregory, D. and Urry, J. (eds) *Social Relations and Spatial Structures*. London: Macmillan.

Harvey, D. (1989a) From managerialism to entrepreneurialism: the transformation of governance in late capitalism, *Geografiska Annaler* 71B: 3–17.

Harvey, D. (1989b) *The Condition of Postmodernity: An Enquiry into the Origins of Cultural Change*. Oxford: Blackwell.

Harvey, D. (1993) From space to place and back again: reflections on the condition of postmodernity, in Bird, J. et al. (eds) *Mapping the Futures: Local Cultures, Global Change*. London: Routledge.

Harvey, D. (1996) *Justice, Nature and the Geography of Difference*. Cambridge, MA: Blackwell.

Harvey, D. (1997) The environment of justice, in Merrifield, A. and Swyngedouw, E. (eds) *The Urbanization of Injustices*. New York: New York University Press.

Harvey, D. (2000) *Spaces of Hope*. Berkeley, CA: University of California Press.

Harvey, D. (2003) *Paris: Capital of Modernity*. London: Routledge.

Haythornthwaite, C. and Wellman, B. (2002) Moving the Internet out of cyberspace, in Wellman, B. and Haythornthwaite, C. (eds) *The Internet and Everyday Life*. Oxford: Blackwell.

Heap, C. (2003) The city as a sexual laboratory: the queer heritage of the Chicago School, *Qualitative Sociology* 26(4): 457–487.

Herbert, S. (2000) On ethnography, *Progress in Human Geography* 24: 550–568.

Highmore, B. (2002a) *Everyday Life and Cultural Theory: An Introduction*. London: Routledge.

Highmore, B. (2002b) Streetlife in London: towards a rhythmanalysis of London in the late nineteenth century, *New Formations* 47: 171–193.

Hill, M.R. (1981) Positivism: a hidden philosophy in geography, in Harvey, M.E. and Holly, B.P. (eds) *Themes in Geographic Thought*. London: Croom Helm.

Hinchliffe, S. (1999) Cities and natures: intimate strangers, in Allen, J., Massey, D. and Pryke, M. (eds) *Unruly Cities?* London: Routledge and Open University.

Hinchliffe, S., Kearnes, M., Degen, M. and Whatmore, S. (2005) Urban wild things: a cosmopolitical experiment, *Environment and Planning D: Society and Space* 23(5): 643–658.

Hochschild, A.R. (2001) *The Time Bind: When Work Becomes Home and Home Becomes Work*. New York: Metropolitan Books.

Hoggart, K. (1991) Let's do away with the rural, *Journal of Rural Studies* 6(3): 245–257.

Hoggart, K. (2004) Walking as an aesthetic practice and a critical tool: some psychogeographical experiments, *Journal of Geography in Higher Education* 28(3): 397–410.

Hollier, D. (1994) Surrealist precipitates: shadows don't cast shadows, *October* 69: 111–132.

Holloway, L. (2002) Smallholding, hobby-farming and commercial farming: ethical identities and the production of farming spaces, *Environment and Planning A* 34: 2055–2070.

Holloway, L. (2004) Donna Haraway, in Hubbard, P., Kitchin, R. and Valentine, G. (eds) *Key Thinkers on Space and Place*. London: Sage.

Holloway, L. and Hubbard, P. (2001) *People and Place: The Extraordinary Geographies of Everyday Life*. London: Prentice Hall.

Holloway, S. (2005) Articulating Otherness? White residents talk about Gypsy-travellers, *Transactions, Institute of British Geographers* 30(3) 351–367.

Holston, J. (1991) *The Modernist City: An Anthropological Critique of Brasilia*. Chicago: Aldine Press.

Hong Kong Special Administrative Region (2000) *Bringing the Vision to Life: Hong Kong's Long-Term Development Needs and Goals*. Hong Kong Central Policy Unit.

Hopkins, J.S.P. (1990) West Edmonton Mall: landscapes of myths and elsewhereness, *Canadian Geographer* 34(1): 2–17.

Hubbard, P. (1996) Urban design and city regeneration: social representations of entrepreneurial landscapes, *Urban Studies* 33(11): 1441–1461.

Hubbard, P. (1997) Sexuality, immorality and the city: red-light districts and the marginalisation of street prostitution, *Gender, Place and Culture* 3(1): 55–72.

Hubbard, P. (2000) Desire/disgust: mapping the moral contours of heterosexuality, *Progress in Human Geography* 24(2): 191–217.

Hubbard, P. (2002) Screen-shifting: consumption, riskless risks and the changing geographies of cinema, *Environment and Planning A* 34(10): 1239–1258.

Hubbard, P. (2004) Revenge and injustice in the neoliberal city: uncovering masculinist agendas, *Antipode* 36(4): 665–682.

Hubbard, P. and Hall, T. (1998) Introducing the entrepreneurial city, in Hall, T. and Hubbard, P. (eds) *The Entrepreneurial City*. Chichester: John Wiley.

Hubbard, P. and Lilley, K. (2004) P(l)acemaking the modern city, *Environment and Planning D: Society and Space* 22(5): 273–294.

Hubbard, P., Kitchin, R., Bartley, B. and Fuller, D. (2002) *Thinking Geographically: Space, Theory and Contemporary Human Geography*. London: Continuum.

Jackson, P. (1985) Urban ethnography, *Progress in Human Geography* 9: 199–213.

Jackson, P. (1989) *Maps of Meaning*. London: Routledge.

Jackson, P. (2000) Rematerializing social and cultural geography, *Social and Cultural Geography* 1: 9–14.

Jackson, P. (2005) The eclipse of urban geography, *Urban Geography* 26(1): 1–3.

Jackson, R.H. (1990) *Cultural Geography: People, Places and Environment*. St Paul, MN: West.

Jacobs, J.M. (1994) Negotiating the heart: 'heritage', redevelopment and identity, *Environment and Planning D: Society and Space* 12(5): 751–772.

Jameson, F. (1984) Post-modernism or the cultural logic of late capitalism, *New Left Review* 146: 53–92.

Jarvis, H., Pratt, A. and Cheng-Chong Wu, P. (2001) *The Secret Life of Cities: The Social Reproduction of Everyday Life*. Harlow: Pearson.

Jayne, M. (2005) Creative cities: the regional dimension, *Environment and Planning C: Government and Policy* 23(4): 537–556.

Jencks, C. (1996) The city that never sleeps, *New Statesman* 28 June: 26–28.

Jenks, C. and Neves, T. (2000) A walk on the wild side: urban ethnography meets the flâneur, *Cultural Values* 4: 1–17.

Jessop, B. (1998) The narrative of enterprise and the enterprise of narrative: place marketing in the entrepreneurial city, in Hall, T. and Hubbard, P. (eds) *The Entrepreneurial City*. Chichester: John Wiley.

Johnston, R.J. (1991) *Geography and Geographers*. London: Arnold.

Johnston, R.J. (2000) Urban geography, in Johnston, R.J., Gregory, D., Pratt, G. and Watts, M. (eds) *Dictionary of Human Geography*. Oxford: Blackwell.

Johnston, R.J., Poulson, M. and Forrest, J. (2006) Modern and post-modern cities and ethnic residential segregation: is Los Angeles different?, *Geoforum* 37 forthcoming.

Jones, J.P. III. and Natter, W. (1997) Identity, space, and other uncertainties, in Benko, G. and Strohmayer, U. (eds) *Space and Social Theory: Geographical Interpretations of Postmodernity*. Oxford: Blackwell.

Jones, J.P. III, Mangieri, T., McCourt, M., Moore, S., Park, K., Pryce-Jones, M. and Woodward, K. (2005) *David Harvey Live*. New York: Continuum.

Kaika, M. (2004) Interrogating the geographies of the familiar: domesticating nature and constructing the autonomy of the modern home, *International Journal of Urban and Regional Research* 28(2): 265–286.

Kaika, M. and Swyngedouw, E. (2000) Fetishizing the modern city: the phantasmagoria of urban technological networks, *International Journal of Urban and Regional Research* 24(1): 120–138.

Katz, P. (1994) *The New Urbanism: Toward an Architecture of Community*. New York: McGraw-Hill.

Kavaratzis, M. and Ashworth, G.J. (2005) City branding: an effective assertion of identity or a transitory marketing trick?, *Tijdschrift voor Economische en Sociale Geografie* 96(5): 506–514.

Keith, M. (2005) *After the Cosmopolitan? Multiculturalism and the Future of Racism*. London: Routledge.

Kenney, J. (1995) Climate, race, and imperial authority: the symbolic landscape of the British Hill Station in India, *Annals of the Association of American Geographers* 85: 694–714.

Kern, S. (1983) *The Culture of Time and Space 1880–1918*. Cambridge, MA: Harvard University Press.

King, A. (1990) *Global Cities: Post-Imperialism and the Internationalization of London*. London: Routledge.

King, A. (2004) *Spaces of Global Cultures*. London: Routledge.

Kinsman, P. (1995) Landscape, race and national identity: the photography of Ingrid Pollard, *Area* 27: 300–310.

Kipfer, S. and Keil, R. (2002) Toronto Inc? Planning the competitive city in the New Toronto, *Antipode* 34(2): 227–264.

Kitchin, R. (1998) *Cyberspace*. Chichester: John Wiley.

Kitchin, R. and Kneale, J. (2001) Science fiction or future fact?, in Kitchin, R. and Kneale, J. (eds) *Lost in Space: Geographies of Science Fiction*. London: Continuum.

Kitchin, R. and Law, R. (2001) The socio-spatial construction of inaccessible public toilets, *Urban Studies* 38(3): 287–298.

Kloosterman, R.C. and Lambregts, B. (2001) Clustering of economic activities in polycentric urban regions: the case of the Randstad, *Urban Studies* 38: 717–732.

Kloosterman, R.C. and Musterd, S. (2001) The polycentric urban region: towards a research agenda, *Urban Studies* 20(6): 623–639

Knox, P. (1991) The restless urban landscape: economic and sociocultural change and the transformation of metropolitan Washington, D.C., *Annals of the Association of American Geographers* 8(1): 181–209.

Knox, P. (1993) *The Restless Urban Landscape*. Englewood Cliffs, NJ: Prentice Hall.

Knox, P. (1995) World cities and the organisation of global space, in Johnston, R., Taylor, P. and Watts, M. (eds) *Geographies of Global Change*. Oxford: Blackwell.

Knox, P. and Pinch, S. (2001) *Urban Social Geography*. Harlow: Pearsons.

Koskela, H. (2000) Video-surveillance and the changing nature of urban space, *Progress in Human Geography* 24(2): 243–265.

Kostof, S. (2001a) *The City Assembled: The Elements of Urban Form through History*. London: Thames and Hudson.

Kostof, S. (2001b) *The City Shaped: Urban Patterns and Meanings through History*. London: Thames and Hudson.

Kracaeur, S. (1927) *The Mass Ornament*. Frankfurt: Suhrkamp.

Krätke, S. (2002) Network analysis of production clusters: the Potsdam/Babelsberg film industry as an example, *European Planning Studies* 10(1): 27–54.

Krätke, S. (2004) City of talents? Berlin's regional economy, socio-spatial fabric and worst practice urban governance, *International Journal of Urban and Regional Research* 28(3): 511–529.

Kresl, P.K. (1995) The determinants of urban competitiveness: a survey, in Kresl, P.K. and Gaert, G. (eds) *North American Cities and the Global Economy*. London: Sage.

Krugman, P. (1998) What's new about the new economic geography?, *Oxford Review of Economic Policy* 14(2): 7–17.

Krutnik, F. (1991) *In a Lonely Street: Film Noir, Genre, Masculinity*. London: Routledge.

Lake, R.W. (2005) Urban crisis redux, *Urban Geography* 26(3): 266–270.

Lambooy, J.G. (1998) Polynucleation and economic development: the Randstad, *European Planning Studies* 6(4): 457–466.

Landry, C. (2000) *The Creative City*. London: Earthscan.

Landry, C., Bianchini, F. and Trotter-Landry, M. (2004) *A Taste of Leicester: Results of an Attitudinal Survey*. Leicester: Comedia.

Lassen, C. (2006) Aeromobility and work, *Environment and Planning A* 38(2): 301–312.

Latham, A. (2003) Urbanity, lifestyle and making sense of the new urban cultural economy, *Urban Studies* 40(9): 1699–1724.

Latham, A. (2004) Ed Soja, in Hubbard, P., Kitchin, R. and Valentine, G. (eds) *Key Thinkers on Space and Place*. London: Sage.

Latham, A. and McCormack, D. (2004) Moving cities: rethinking the materialities of urban geographies, *Progress in Human Geography* 28(6): 701–724.

Latour, B. (1987) *Science in Action: How to Follow Engineers in Society*. Milton Keynes: Open University Press.

Latour, B. (1993) *We Have Never been Modern*. London: Harvester Wheatsheaf.

Latour, B. (2003) Is remodernization occurring – and if so, how to prove it? A commentary on Ulrich Beck, *Theory, Culture and Society* 20(1): 35–48.

Latour, B. and Hermant, E. (2001) *Paris: ville invisible*. Paris: La Découverte-Les Empêcheurs de Penser en Rond.

Lauria, M. and Knopp, L. (1985) Towards an analysis of the role of gay communities in the urban renaissance, *Urban Geography* 6: 152–169.

Laurier, E. (1998) Geographies of talk: 'Max left a message for you', *Area* 31(1): 35–48.

Laurier, E. (2001) Why people say where they are on mobile phones, *Environment and Planning D: Society and Space* 19: 485–504.

Laurier, E. (2004) Paul Virilio, in Hubbard, P., Kitchin, R. and Valentine, G. (eds) *Key Thinkers on Space and Place*. London: Sage.

Laurier, E., Brown, B. and Lorimer, H. (2005) *Habitable Cars: What We Do There.* Online papers archived by the Institute of Geography, School of Geosciences, University of Edinburgh http://hdl.handle.net/1842/815.

Law, J. (1992) Notes on the theory of the actor-network, *Systems Practice* 5: 379–393.

Law, J. (1999) After ANT: complexity, naming and topology, in Law, J. and Hassard, J. (eds) *Actor Network Theory and After*. Oxford: Blackwell.

Law, J. and Hassard, J. (eds) (1999) *Actor Network Theory and After*. Oxford: Blackwell.

Lees, L. (2001) Towards a critical geography of architecture: the case of an ersatz colosseum, *Ecumene: A Journal of Cultural Geographies* 8(1): 51–86.

Lees, L. (2003) The ambivalence of diversity and the politics of urban renaissance: the case of youth in downtown Portland, Maine, USA, *International Journal of Urban and Regional Research* 27(3): 613–634.

Lees, L. and Demeritt, D. (1998) Envisioning the liveable city, *Urban Geography* 19(5): 332–359.

Lefebvre, H. (1972) *La Pensée marxiste et la ville*. Paris: Casterman.

Lefebvre, H. (1991) *The Production of Space*. Cambridge, MA: Blackwell.

Lefebvre, H. (1995) *Introduction to Modernity*. London: Verso.

Lefebvre, H. (1996) *Writings on Cities*. Oxford: Blackwell.

LeGates, R.T. and Stout, F. (eds) (1996) *The City Reader*. New York: Routledge.

Leitner, H. and Sheppard, E. (1998) Economic uncertainty, inter-urban competition and the efficacy of entrepreneurialism, in Hall, T. and Hubbard, P. (eds) *The Entrepreneurial City*. Chichester: John Wiley.

Leitner, H. and Sheppard, E. (2002) The city is dead, long live the network, *Antipode* 34(3): 495–518.

Ley, D. (1991) Co-operative housing as a moral landscape: re-examining the postmodern city, in Duncan, J. and Ley, D. (eds) *Place/Culture/Representation*. London: Routledge.

Ley, D. (1996) *Gentrification and the Middle Classes*. Oxford: Oxford University Press.

Ley, D. (2003) Artists, aestheticisation and the field of gentrification, *Urban Studies* 40: 2527–2544.

Leyshon, A., Matless, D. and Revill, G. (eds) (1998) *The Place of Music*. New York: Guilford.

Lilley, K.D. (2002) *Urban Life in the Middle Ages: 1000–1450AD*. Basingstoke: Palgrave.

Llewellyn, M. (2003) Polyvocalism and the public: 'doing' a critical historical geography of architecture, *Area* 35(3): 264–270.

Lloyd, R. (2002) Neo-bohemia: art and neighbourhood redevelopment in Chicago, *Journal of Urban Affairs* 24(5): 517–532.

Lofland, L. (1998) *The Public Realm*. New York: Aldine.

Loftman, P. and Nevin, B. (1998) Going for growth: prestige projects in three British cities, *Urban Studies* 33(6): 991–1019.

Loftman, P. and Nevin, B. (2003) Prestige projects, city centre restructuring and social exclusion: taking the long-term view, in Miles, M. and Hall, T. (eds) *Urban Futures: Critical Commentaries on Shaping the City*. London: Routledge.

Logan, J.R. and Molotch, H. (1987) *Urban Fortunes: The Political Economy of Place*. Berkeley, CA: University of California Press.

Longhurst, R. (2000) 'Corporeographies' of pregnancy: 'bikini babes', *Environment and Planning D: Society and Space* 18: 453–472.

Longhurst, R. (2001) *Bodyspace*. London: Routledge.

Lorimer, H. (2005) Cultural geography: the busyness of being more than representational, *Progress in Human Geography* 29(1): 83–94.

Lowenthal, D. (1991) British national identity and the English landscape, *Rural Studies* 2(2): 205–230.

Luke, T. and O'Tuathail, G. (2000) Thinking geopolitical space: the spatiality of war, speed and vision, in Crang, M. and Thrift, N. (eds) *Thinking Space*. London: Routledge.

Lupton, D. (1999) Monsters in metal cocoons: 'road rage' and cyborg bodies, *Body and Society* 5: 57–72.

Lynch, K. (1960) *The Image of the City*. Cambridge, MA: MIT Press.

Lyon, D. (2002) Surveillance studies: understanding visibility, mobility and the phenetic fix, *Surveillance and Society* 1: 1–7, http://www.surveillance-and-society.org/articles1/editorial.pdf

Lyon, D. (2003) *Surveillance after September 11th*. Cambridge: Polity.

Lyon Cite (1998) *Charte Lyon – Marseille*. Lyon: Cite Lyon.

McCann, E. (1995) Neotraditional developments: the anatomy of a new urban form, *Urban Geography* 16(3): 210–233.

McCann, E. (1999) Race, protest, and public space: contextualizing Lefebvre in the US city, *Antipode* 31(2): 163–184.

McCann, E. (2002) The urban as an object of study in Global Cities literature: representational practices and questions of scale, in Herod, A. and Wright, M. (eds) *Geographies of Power: Placing Scale*. Cambridge, MA: MIT Press.

McCann, E. (2004a) 'Best places': interurban competition, quality of life and popular media discourse, *Urban Studies* 41: 1909–1929.

McCann, E. (2004b) Urban political economy beyond the global city, *Urban Studies* 41: 2315–2333.

McDowell, L. (1999) *Gender, Identity and Place: Understanding Feminist Geographies*. Cambridge: Polity.

MacKenzie, A. (2005) The performativity of code: software and cultures of circulation, *Theory, Culture and Society* 22(1): 71–92.

McKenzie, J. (2001) *Perform or Else: From Discipline to Performance*. London: Routledge.

MacLeod, G. (1999) Place, Politics and 'scale dependence': exploring the structuration of Euro-regionalism, *European Urban and Regional Studies* 6(3): 231–254.

MacLeod, G. and Goodwin, M. (1999) Space, scale and state strategy: towards a rethinking of urban and regional governance, *Progress in Human Geography* 23(4): 503–527.

MacLeod, G. and Ward, K. (2002) Spaces of utopia and dystopia: landscaping the contemporary city, *Geografiska Annaler* 84B: 153–170.

McLuhan, M. (1959) *Understanding Media*. New York: Basic Books.

McNeill, D. (2005) Skyscraper geography, *Progress in Human Geography* 29(1): 41–55.

Malbon, B. (1999) *Clubbing: Clubbing Cultures and Experience*. London: Routledge.

Malmberg, A. and Maskell, P. (2002) The elusive concept of localization economies: towards a knowledge-based theory of spatial clustering, *Environment and Planning A* 34(3): 429–449.

Markusen, A. (2005) Urban development and the politics of a creative class: evidence from the study of artists, unpublished paper available at www.hhh.umn.edu/projects/prie/

Marston, S.A. and Manning, K.E. (2005) Teaching the cultural politics and economy of global cities in the periphery, *Urban Geography* 26 (3) 252–256.

Marx, K. (1867) [1887] *Capital – Volume One*. Moscow: Progress.

Massey, D. (1991a) A global sense of place, *Marxism Today* June: 24–29.

Massey, D. (1991b) Flexible sexism, *Environment and Planning D: Society and Space* 9: 31–58.

Massey, D. (1994) *Space, Place, and Gender*. Minneapolis, MN: University of Minnesota Press.

Massey, D. (1999) Cities in the world, in Massey, D., Allen, J. and Pile, S. (eds) *City Worlds*. London: Routledge.

Massey, D., Allen, J. and Pile, S. (eds) (1999) *City Worlds*. London: Routledge.

Matless, D. (1998) *Landscape and Englishness*. London: Reaktion.

Matless, D. (2000) Five objects, in Naylor, S., Ryan, J. and Cook, I. (eds) *Cultural Turns / Geographical Turns*. London: Pearson.

May, T. and Perry, B. (2005) Continuities and change in urban sociology, *Sociology* 39(2): 343–370.

Means, M. and Tims, C. (2005) *People-making Places: Growing the Public Life of Cities*. London: Demos.

Merleau-Ponty, M. (1962) *Phenomenology of Perception*. London: Routledge & Kegan Paul.

Merrifield, A. (1995) Lefebvre, anti-logos and Nietzsche: an alternative reading of *The Production of Space*, *Antipode* 27(3): 294–303.

Merrifield, A. (2002) *Metromarxism*. New York: Routledge.

Merrifield, A. (2005) The sentimental city: the lost urbanism of Pierre Orlan and Guy Debord, *International Journal of Urban and Regional Research* 28(4): 930–940.

Merriman, P. (2004) Driving places: Marc Augé, non-places and the geographies of the M1 motorway, *Theory, Culture and Society* 21(6): 145–167.

Merriman, P. (2005) Materiality, subjectification and government: the geographies of Britain's Motorway Code, *Environment and Planning D: Society and Space* 23(2): 235–250.

Metcalf, B. (1996) *Making Muslim Space in North America and Europe*. Berkeley, CA: University of California Press.

Miles, M. (1998) *Art, Space and the City*. London: Routledge.

Miles, M. (2002) Wish you were here, in Speir, S. (ed.) *Urban Visions: Experiencing and Envisioning the City*. Liverpool: Tate Liverpool Press.

Milgram, S. (1970) Experience of living in cities, *Science* 167: 1461–1468.

Milicevic, A.S. (2001) Radical intellectuals: whatever happened to the new urban sociology, *International Journal of Urban and Regional Research* 25(4): 769–785.

Miller, D. (1998) Why some things matter, in Miller, D. (ed.) *Material Culture: Why Some Things Matter*. Chicago, IL: University of Chicago Press.

Mills, C. (1993) Myths and meanings of gentrification, in Duncan, J. and Ley, D. (eds) *Place/Culture/Representation*. London: Routledge.

Mitchell, D. (1997) The annihilation of space by law: the roots and implications of anti-homeless laws, *Antipode* 29: 303–335.

Mitchell, D. (2003) *The Right to the City*. New York: Guilford.

Mitchell, W.J. (1995) *City of Bits: Space, Place, and the Infobahn*. Cambridge, MA: MIT Press.

Mitchell, W.J. (2004) *Me++ The Cyborg City and the Networked City*. Cambridge, MA: MIT Press.

Molotch, H. (1996) L.A. as a design product: how design works in a regional economy, in Scott, A.J. and Soja, E. (eds) *The City: Los Angeles and Urban Theory at the End of the Twentieth Century*, Berkeley, CA: University of California Press.

Molotch, H. (2003) *Where Stuff Comes From: How Toasters, Toilets, Cars, Computers, and Many Other Things Come to Be as They Are*. London: Routledge.

Morgan, K. (2004) The exaggerated death of geography: learning, proximity and territorial innovation systems, *Journal of Economic Geography* 4: 3–21.

Mumford, L. (1961) *The City in History: Its Origins, its Transformations, and its Prospects*. New York: Harcourt, Brace and World.

Murdoch, J. (1995) Actor-networks and the evolution of economic forms: combining description and exploration in theories of regulation, flexible specialization, and networks, *Environment and Planning A* 27: 731–757.

Murdoch, J. (1997) Inhuman/nonhuman: actor-network theory and the prospects for a nondualistic and symmetrical perspective on nature and society, *Environment and Planning D: Society and Space* 15(6): 731–756.

Murdoch, J. (1998) The spaces of actor network theory, *Geoforum* 29(4): 357–384.

Murdoch, J. (2003) Introducing hybridity, in Cloke, P. (ed.) *Country Visions*. Harlow: Pearson.

Murdoch, J. (2004) Humanising posthumanism, *Environment and Planning D: Society and Space* 36: 1356–1360.

Murdoch, J. and Marsden, T. (1995) The spatialisation of politics: local and national actor spaces in environmental conflict, *Transactions, Institute of British Geographers* 20: 368–380.

Musterd, S. and van Zelm, I. (2001) Polycentricity, households and the identity of places, *Urban Studies* 38(4): 679–696.

Myslik, W. (1996) Renegotiating the social/sexual identities of places: gay communities as safe havens or sites of resistance?, in Duncan, N. (ed.) *Body Space: Destabilising Geographies of Gender and Sexuality*. London: Routledge.

Nachum, L. and Keeble, D. (2003) Neo-Marshallian clusters and global networks: the linkages of media firms in Central London, *Long Range Planning* 36(5): 459–480.

Nash, C. (1993) 'Embodying the nation': the West of Ireland landscape and Irish identity, in O'Connor, B. and Cronin, M. (eds) *Tourism in Ireland: A Critical Analysis*. Cork: Cork University Press.

Nash, C. (1996) Reclaiming vision: looking at landscape and the body, *Gender, Place and Culture: A Journal of Feminist Geography* 3(2): 149–169.

Nash, C. (2000) Performativity in practice: some recent work in cultural geography, *Progress in Human Geography* 24: 653–664.

Nead, L. (2000) *Victorian Babylon: People, Streets and Images in Nineteenth Century London*. New Haven, CT: Yale University Press.

Neff, G. (2005) The changing place of cultural production: social networks in the digital media industry, *Annals of the American Academy for Political and Social Science* 597: 134–152.

Neumann, R.P. (1995) Ways of seeing Africa: colonial recasting of African society and landscape in Serengeti national park, *Ecumene* 2(2): 149–169.

Nochlin, L. (1971) *Realism*. Harmondsworth: Penguin.

Norris, C. and Armstrong, G. (1999) *The Maximum Surveillance Society: The Rise of CCTV*. Oxford: Berg.

Nye, D. (1990) *Electrifying America: Social Meanings of a New Technology 1880–1940*. Cambridge, MA: MIT Press.

OECD (2002) *OECD Information Technology Outlook*. Paris: OECD.

Ogborn, M. (1998) *Spaces of Modernity: London's Geographies 1680–1780*. New York: Guilford.

Pacione, M. (1999) Applied geography: in pursuit of useful knowledge, *Applied Geography* 19(1): 1–12.

Pahl, R.E. (1970) *Whose City? And Other Essays on Sociology and Planning*. Harlow: Longman.

Pahl, R.E. (1977) *Readings in Urban Sociology*. London: Routledge.

Park, R., Burgess, E. and McKenzie, R. (eds) (1925) *The City*. Chicago, IL: University of Chicago Press.

Parr, J.B. (2002) Agglomeration economies: ambiguities and confusions, *Environment and Planning A* 34(6): 717–731.

Parr, J. (2004) The polycentric urban region: a closer inspection, *Regional Studies* 38(3): 231–240.

Patch, J. (2004) The embedded landscape of gentrification, *Visual Studies* 19(2): 169–186.

Peck, J. (2001) Neoliberalizing states: thin policies/hard outcomes, *Progress in Human Geography* 25(3): 445–455.

Peck, J. (2005) Struggling with the creative class, *International Journal of Urban and Regional Research* 29(4): 740–770.

Peet, R. (1998) *Modern Geographical Thought*. Oxford: Blackwell.

Peet, R. (2005) Bio-gaze, *Environment and Planning A* 37(1): 161–187.

Perl, J. (2004) *New Art City: Manhattan at Mid-century*. New York: Knopf.

Phelps, N.A. (2004) Clusters, dispersion and the spaces in between: for an economic geography of the banal, *Urban Studies* 41(5–6): 971–989.

Phillips, R. (1999) Travelling sexualities: Richard Burton's sotadic zone, in Duncan, J. and Gregory, D. (eds) *Writes of Passage: Reading Travel Writing*. London: Routledge.

Phillips, R., West, D. and Shuttleton, D. (eds) (2000) *Decentering Sexualities: Politics and Representations Beyond the Metropolis*. London: Routledge.

Philo, C. (1995) Animals, geography and the city: notes on inclusions and exclusions, *Environment and Planning D: Society and Space* 13(6): 655–681.

Philo, C. (2000) More words, more worlds: reflections on the 'cultural turn' and human geography, in Cook, I., Crouch, D., Naylor, S. and Ryan, J. (eds) *Cultural Turns/Geographical Turns*. Harlow: Prentice Hall.

Pike, D. (2002) Modernist space and the transformation of the London Underground, in Gilbert, P. (ed.) *Imagined Londons*. Albany, NY: State University of New York Press.

Pile, S. (1999) What is a city?, in Massey, D., Allen, J. and Pile, S. (eds) *City Worlds*. London: Routledge.

Pile, S. (2005) *Real Cities*. London: Sage.

Pinder, D. (1996) Subverting cartography, *Environment and Planning A* 28: 405–427.

Pinder, D. (2000) Old Paris is no more: geographies of spectacle and anti-spectacle, *Antipode* 32: 357–386.

Pinder, D. (2004) Inventing new games: unitary urbanism and the politics of space, in Lees, L. (ed.) *The Emancipatory City: Paradoxes and Possibilities*. London: Sage.

Pirenne, H. (1925) *Medieval Cities: Their Origins and the Revival of Trade*. Princeton, NJ: Princeton University Press.

Plant, S. (1992) *The Most Radical Gesture: The Situationist International in a Postmodern Age*. London: Routledge.

Pocock, D. (1981) *Humanistic Geography and Literature: Essays on the Experience of Place*. London: Croom Helm.

Poovey, M. (1995) *Making a Social Body: British Cultural Formation, 1830–1864*. Chicago, IL: University of Chicago Press.

Porter, M. (1998) Clusters in the new economics of competition, *Harvard Business Review* 76(6): 77–90.

Power, D. and Jansson, J. (2004) The emergence of a post-industrial music economy? Music and ICT synergies in Stockholm, Sweden, *Geoforum* 35(4): 425–439.

Pratt, A. (2004) The cultural economy: a call for spatialised production of culture perspectives, *International Journal of Cultural Studies* 7(1): 117–128.

Pratt, G. (1991) Feminist analyses of the restructuring of urban life, *Urban Geography* 12(5): 594–605.

Pratt, G. and San Juan, R.M. (2004) In search of the horizon: utopia in *The Truman Show* and *The Matrix*, in Lees, L. (ed.) *The Emancipatory City: Paradoxes and Possibilities*. London: Sage.

Randels, S. and Dicken, P. (2004) 'Scale' and the instituted construction of the urban: contrasting the cases of Manchester and Lyon, *Environment and Planning A* 36(11): 2011–2032.

Reckless, W.C. (1926) *The Distribution of Commercialized Vice in the City: A Sociological Analysis*. Chicago, IL: Publications of American Sociological Society.

Relph, E. (1976) *Place and Placelessness*. London: Pion.

Rendell, J. (2002) *The Pursuit of Pleasure*. London: Athlone.

Rex, J.A. and Moore, R. (1974) *Race, Conflict and Community: A Case Study of Sparkbrook, Birmingham*. London: Oxford University Press.

Ritzer, G. (1999) *Enchanting a Disenchanted World: Revolutionizing the Means of Consumption*. Los Angeles, CA: Pine Forge Press.

Roberts, G.K. and Steadman, P. (eds) (1999) *American Cities and Technology Reader: Wilderness to Wired City*. London: Routledge.

Roberts, J. (1999) Philosophising the everyday, *Radical Philosophy* 98: 12–17.

Robins, K. (1995) The new spaces of global media, in Johnston, R.J., Taylor, P.J. and Watts, M.J. (eds) *Geographies of Global Change: Remapping the World in the Late Twentieth Century*. Oxford: Basil Blackwell.

Robins, K. and Webster, F. (1988) Athens without slaves . . . or slaves without Athens? The neurosis of technology, *Science as Culture* 3(1): 7–53.

Robinson, J. (2002) Global and world cities: a view from off the map, *International Journal of Urban and Regional Research* 26: 531–534.

Robinson, J. (2006) *Ordinary Cities: Between Modernity and Development*. London: Routledge.

Rojek, C. (2005) The consumerist syndrome: an interview with Zygmunt Bauman, *Journal of Consumer Culture* 4(3): 291–312.

Rose, G. (1997) Engendering the slum: photography in east London in the 1930s, *Gender, Place and Culture* 4(3): 277–300.

Rose, G. (2001) *Visual Methodologies*. London: Sage.

Rose, G. (2003) On the need to ask how, exactly, is geography visual?, *Antipode* 35(2): 212–221.

Ross, K. (1994) *Fast Cars, Clean Bodies*. New York: Verso.

Ruskin, J. (1880) *Letters to the Clergy on the Lord's Prayer and the Church*. London: George Allen.

Sadler, S. (1998) *The Situationist City*. Cambridge, MA: MIT Press.

Sandercock, L. (2003) *Cosmopolis II: Mongrel Cities of the 21st Century*. London: Continuum.

Santagata, W. (2002) Cultural districts, property rights and sustainable economic growth, *International Journal of Urban and Regional Research* 26(1): 9–23.

Sassen, S. (1991) *The Global City: New York, London, Tokyo*. Princeton, NJ: Princeton University Press.

Sassen, S. (1996) Whose city is it? Globalization and the formation of new claims, *Public Culture* 8: 205–223.

Saunders, P. (1981) *Social Theory and the Urban Question*. London: Hutchinson.

Saunders, P. (1986) *Social Theory and the Urban Question*, 2nd edn. London: Routledge.

Savage, M., Warde, A. and Ward, K. (2003) *Urban Sociology, Capitalism and Modernity*, 2nd edn. Basingstoke: Palgrave Macmillan.

Savitch, H.V. and Kantor, P. (2002) *Cities in the International Marketplace: The Political Economy of Urban Development in North America and Western Europe*. Princeton, NJ: Princeton University Press.

Schilling, C. (2005) *Bodies, Technology and Society*. London: Sage.

Schivelbusch, W. (1986) *The Railway Journey: The Industrialisation of Time and Space in the Nineteenth Century*. Berkeley, CA: University of California Press.

Schivelbusch, W. (1988) *Disenchanted Night: The Industrialisation of Light in the Nineteenth Century*. Berkeley, CA: University of California Press.

Schlör, J. (1998) *Nights in the Big City*. London: Reaktion.

Scott, A. (1980) *The Urban Land Nexus and the State*. London: Pion.

Scott, A. (1988) *New Industrial Spaces: Flexible Production, Organization and Regional Development in North America and Western Europe*. London: Pion.

Scott, A. (1996) The craft, fashion, and cultural-products industries of Los Angeles: competitive dynamics and policy dilemmas in a multisectoral image-producing complex, *Annals of the Association of American Geographers* 86: 306–323.

Scott, A. (1999) The US recorded music industry: on the relations between organization, location and creativity in the cultural economy, *Environment and Planning A* 31(11): 1965–1984.

Scott, A. (2000) *The Cultural Economy of Cities*. London: Sage.

Scott, A. (2001) Capitalism, cities, and the production of symbolic forms, *Transactions of the Institute of British Geographers* 26: 11–22.

Scott, A. (2002) A new map of Hollywood: the production and distribution of American motion pictures, *Regional Studies* 36: 957–975.

Scott, A. and Soja, E. (eds) (1996) *The City: Los Angeles and Urban Theory and the End of the Twentieth Century*. Berkeley, CA: University of California Press.

Scott, A. and Storper, M. (1992) *Pathways to Industrialisation*. London: Routledge.

Scott, J.C. (1998) *Seeing Like a State: How Certain Schemes to Improve the Human Condition have Failed*. New Haven, CT: Yale University Press.

Seamon, D. (1979) *A Geography of the Lifeworld*. New York: St Martin's Press.

Sennett, R. (1994) *Flesh and Stone: The Body and the City in Western Civilization*. London: Faber & Faber.

Sharp, J.P., Routledge, P., Philo, C. and Paddison, R. (2000) Entanglements of power: geographies of domination/resistance, in Sharp, J.P., Routledge, P., Philo, C. and Paddison, R. (eds) *Entanglements of Power*. London: Routledge.

Sheller, M. and Urry, J. (2006) The new mobilities paradigm, *Environment and Planning A* 38(2): 207–226.

Shevky, E. and Bell, W. (1955) *Social Area Analysis*. Stanford, CA: Stanford University Press.

Shields, R. (1991) *Places on the Margin*. London: Routledge.

Shields, R. (1997) Ethnography in the crowd: the body, sociality and globalization in Seoul, *Focaal, Tijdschrift voor Antropologie* 30–31: 7–21.

Shields, R. (1999) *Lefebvre, Love and Struggle*. London: Routledge.

Short, J.R. (1991) *Imagined Country: Environment, Culture, Society*. London: Routledge.

Short, J.R. (1996) *The Urban Order*. Oxford: Blackwell.

Short, J.R. (2004) *Global Metropolitan: Globalizing Cities in a Capitalist World*. New York: Routledge.

Short, J.R. and Kim, Y-H. (1998) Urban crises/urban representations: selling the city in difficult times, in Hall, T. and Hubbard, P. (eds) *The Entrepreneurial City*. Chichester: John Wiley.

Short, J.R. and Kim, Y.H. (1999) *Globalization and the City*. Harlow: Longman.

Short, J.R., Benton, L.M., Luce, W.B. and Walton, J. (1993) Reconstructing the image of an industrial city, *Annals of the Association of American Geographers* 83(2): 207–224.

Short, J.R., Kim, Y-H., Kuus, M. and Wells, H. (1996) The dirty little secret of world city research: data problems in comparative analysis, *International Journal of Urban and Regional Research* 20(4): 697–719.

Shurmer-Smith, P. and Hannam, K. (1994) *Worlds of Desire, Realms of Power*. London: Arnold.

Sibley, D. (1995) *Geographies of Exclusion: Society and Difference in the West*. London: Routledge.

Sibley, D. (2001) The binary city, *Urban Studies* 38(2): 239–250.

Sidaway, J. (2000) Postcolonial geographies: an exploratory essay, *Progress in Human Geography* 24(4): 591–612.

Simmel, G. (1950) The metropolis and mental life, in Wolff, K. (ed.) *The Sociology of Georg Simmel*. New York: Free Press.

Simonsen, K. (2005) Bodies, sensations, space and time: the contribution from Henri Lefebvre, *Geografiska Annaler* 87B(1): 1–14.

Simpson-Housley, P. and Preston, P. (1994) *Writing the City: Eden, Babylon and the New Jerusalem*. London: Routledge.

Sinclair, I. (1997) *Lights Out for the Territory*. London: Granta.

Sinclair, I. (2003) *London Orbital*. London: Granta.

Sklair, L. (2005) The transnational capitalist class and contemporary architecture in globalising cities, *International Journal of Urban and Regional Research* 29(3): 485–500.

Slater, D. (1998) Semiotics and content analysis, in Seal, C. (ed.) *Researching Society and Culture*. London: Sage.

Slater, T. (2002) Fear of the city 1882–1967: Edward Hopper and the discourse of anti-urbanism, *Social and Cultural Geography* 3(2): 134–154.

Smith, D.A. and Timberlake, M. (1995) Conceptualizing and mapping the structure of the world's city system, *Urban Studies* 32: 287–302.

Smith, D.M. (1975) *Human Geography: A Welfare Approach*. London: Arnold.

Smith, N. (1979) Towards a theory of gentrification: a back to the city movement of capital not people, *Journal of the American Planning Association* 45: 538–548.

Smith, N. (1986) Gentrification, the frontier, and the restructuring of urban space, in Smith, N. and Williams, P. (eds) *Gentrification and the City*. Boston, MA: Allen & Unwin.

Smith, N. (1996) *The New Urban Frontier: Gentrification and the Revanchist City*. London: Routledge.

Smith, R.G. (2003) World city topologies, *Progress in Human Geography* 27(5): 561–582.

Soja, E. (1989) *Postmodern Geographies: The Reassertion of Space in Critical Social Theory*. London: Verso.

Soja, E. (1996) *Thirdspace: Journeys to Los Angeles and Other Real-and-Imagined Places*. Oxford: Blackwell.

Soja, E. (2000) *Postmetropolis: Critical Studies of Cities and Regions*. Oxford: Blackwell.

Solnit, R. (2000) Farewell Bohemia: on art, rent and urbanity, *Harvard Design Magazine* 11: 62–69.

Sorkin, M. (1992) 'See you in Disneyland', in Sorkin, M. (ed.) *Variations on a Theme Park: The New American City and the End of Public Space*. New York: Hill and Wang.

Steel, M. and Symes, M. (2005) The privatisation of public space? The American experience of Business Improvement Districts and their relationship to local governance, *Local Government Studies* 31(3): 321–334.

Stone, C. (1989) *Regime Politics: Governing Atlanta 1946–1988*. Lawrence, KS: University of Kansas.

Storper, M. (1993) Regional worlds of production: learning and innovation in the technology districts of France, Italy and the USA, *Regional Studies* 7(3): 433–455.

Storper, M. (1996) The world of the city: local relations in a global economy, mimeo. Los Angeles, CA: School of Public Policy and Social Research, UCLA.

Storper, M. (1997) Territories, flows and hierarchies in the global economy, in Cox, K.R. (ed.) *Spaces of Globalization*. New York: Guilford.

Storper, M. and Venables, A.J. (2004) Buzz: face-to-face contact and the urban economy, *Journal of Economic Geography* 4(4): 351–370.

Sudjic, D. (1991) *Thousand Mile City*. London: Verso.

Susser, I. (2002) Manuel Castells: conceptualising the city in the information age, in Susser, I. (ed.) *The Castells Reader on Cities and Social Theory*. Oxford: Blackwell.

Swyngedouw, E. (1989) The heart of the place: the resurrection of locality in an age of hyperspace, *Geografiska Annaler B* 71: 31–42.

Swyngedouw, E. (1997) Neither global nor local: 'glocalization' and the politics of scale, in Cox, K. (ed.) *Spaces of Globalization: Reasserting the Power of the Local*. New York: Guilford.

Swyngedouw, E. (1999) Modernity and hybridity: the production of nature, water, modernisation in Spain, *Annals of the Association of American Geographers* 89: 443–465.

Taylor, P.J. (1985) *Political Geography: World-economy, Nation-state and Locality*. London: Longman.

Taylor, P.J. (1999a) Places, spaces and Macy's: place-space tensions in the political geography of modernities, *Progress in Human Geography* 23(1): 7–26.

Taylor, P.J. (1999b) Worlds of large cities: pondering Castells' space of flows, *Third World Planning Review* 21(3): 3–10.

Taylor, P.J. (2001) Urban hinterworlds: geographies of corporate service provision under conditions of contemporary globalization, *Geography* 86(1): 51–60.

Taylor, P.J. (2004) *World City Network*. London: Routledge.

Taylor, P.J. and Lang, R. (2004) The shock of the new: concepts describing recent urban change, *Environment and Planning A* 36: 951–958.

Thrift, N.J. (1980) Behavioural geography, in Wrigley, N. and Bennett, R.J. (eds) *Quantitative Geography: A British View*. London: Routledge & Kegan Paul.

Thrift, N.J. (1993) An urban impasse, *Theory, Culture and Society* 10: 229–238.

Thrift, N.J. (1994) On the social and cultural determinants of international financial centres: the case of the City of London, in Corbridge, S., Martin, R. and Thrift, N.J. (eds) *Money, Power and Space*. Oxford: Blackwell.

Thrift, N. (1995) A hyperactive world?, in Taylor, P., Watts, M. and Johnston, R. (eds) *Geographies of Global Change*. Oxford: Blackwell.

Thrift, N.J. (1996a) New urban eras and old technological fears: reconfiguring the goodwill of electronic things, *Urban Studies* 33: 1463–1494.

Thrift, N.J. (1996b) *Spatial Formations*. London: Sage

Thrift, N.J. (1997) Cities without modernity, cities with magic, *Scottish Geographical Magazine* 113(2): 138–149.

Thrift, N. (1998) Distance is not a safety zone but a field of tension: mobile geographies and world cities, *Nederlandse Geografische Studies* 241: 54–66.

Thrift, N. (1999) Steps towards an ecology of place, in Massey, D., Allen, J. and Sarre, P. (eds) *Human Geography Today*. Cambridge: Polity.

Thrift, N. (2000) Performing cultures in the new economy, *Annals of the Association of American Geographers* 90: 674–692.

Thrift, N. (2003a) Closer to the machine?, *Cultural Geographies* 10: 389–407.

Thrift, N. (2003b) Space: the fundamental stuff of human geography, in Valentine, G., Clifford, N. and Rice, S. (eds) *Key Concepts in Geography*. London: Sage.

Thrift, N. (2004) Driving in the city, *Theory, Culture and Society* 21(6): 41–59.

Thrift, N. (2005a) But malice aforethought: cities and the natural history of hatred, *Transactions, Institute of British Geographers* 30(2): 133–150.

Thrift, N. (2005b) Panicsville: Paul Virilio and the esthetic of disaster, *Cultural Politics* 1(3): 337–348.

Tivers, J. (1985) *Women Attached: The Lives of Young Women with Young Children*. London: Croom Helm.

Tonnies, F. (1887) *Community and Society*. New York: Harper & Row.

Townsend, A.M. (2001) Network cities and the global structure of the Internet, *American Behavioral Scientist* 44(10): 1697–1711.

Transnational Communities Programme (1998) *Programme Brochure*. Oxford: Economic and Social Research Council.

Tufts, S. (2000) Building the 'competitive city': labour and Toronto's bid to host the Olympic games, *Geoforum* 35(1): 47–58.

Urry, J. (2000) *Sociology beyond Societies: Mobilities for the Twenty-first Century*. London: Routledge.

Urry, J. (2001) Transports of delight, *Leisure Studies* 20: 237–245.

Urry, J. (2004) The new mobilities paradigm, unpublished paper.

Valentine, G. (1989) A geography of fear, *Area* 21(4): 385–390.

Valentine, G. (1997) Making space: separatism and difference, in Jones III, J.P., Nast, H.J. and Roberts, S.M. (eds) *Thresholds in Feminist Geography: Difference, Methodology, Representation*. Lanham, MD: Rowman & Littlefield.

Valentine, G. (1999) A corporeal geography of consumption, *Environment and Planning D: Society and Space* 17(3): 329–351.

Van Dijk, M. (1999) The one-dimensional network society of Manuel Castells, *New Media and Society* 1(1): 127–138.

van Weesep, J. (1994) Gentrification as a research frontier, *Progress in Human Geography* 18(1): 74–83.

Vertovec, S. (2004) Cheap calls: the social glue of migrant transnationalism, *Global Networks* 4(2): 1470–2266.

Vidler, A. (1994) Bodies in spaces/subjects in the city: psychopathologies of modern urbanism, *Differences* 5(3): 31–51.

Vidler, A. (2002) Photojournalism: planning the city from above and from below, in Bridge, G. and Watson, S. (eds) *The City Companion*. Oxford: Blackwell.

Vion, A. (2002) Europe from the bottom up: town twinning in France during the Cold War, *Contemporary European History* 11(4): 623–640.

Virilio, P. (1986a) *Speed and Politics*. trans. Polizotti, M. New York: Semiotext(e).

Virilio, P. (1986b) The overexposed city, *Zone* 1: 14–39.

Virilio, P. (1991) *The Lost Dimension*. New York: Semiotext(e).

Virilio, P. (1997) The overexposed city, in Leach, N. (ed.) *Rethinking Architecture.* London: Routledge.

Virilio, P. (1999) Indirect light, *Theory, Culture, Society* 16(5–6): 57–70.

Virilio, P. (2005) *City of Panic.* Oxford: Berg.

Wagner, P. and Mikesell, M. (1962) *Readings in Cultural Geography.* Chicago, IL: University of Chicago Press.

Walby, S. (1997) *Gender Transformation.* London: Routledge.

Wall, M. (1997) Stereotypical constructions of the Maori race in the media, *New Zealand Geographer* 53(2): 40–45.

Wallerstein, I. (1979) *The Capitalist World Economy.* Cambridge: Cambridge University Press.

Walmsley, D.J. (1988) *Urban Living: The Individual in the City.* London: Longman.

Ward, K. (1996) Rereading urban regime theory: a sympathetic critique, *Geoforum* 27: 427–438.

Ward, K. (2000) From rentiers to rantiers: active entrepreneurs, structural speculators and the politics of marketing the city, *Urban Studies* 37(7): 1093–1107.

Ward, K. and Jonas, A. (2004) Competitive city-regionalism as a politics of space: a critical reinterpretation of the new regionalism, *Environment and Planning A* 36: 2119–2139.

Watson, S. (1999) Cities of difference, in Allen, J., Massey, D. and Pryke, M. (eds) *Unruly Cities?* London: Routledge and the Open University.

Watson, S. and Gibson, K. (1995) *Postmodern Cities and Spaces.* Oxford: Blackwell.

Weber, M. (1922) The city, in *Wirtschaft und Gesellschaft*, Tübingen: J.C.B. Moir.

Weber, M. (1923) *Economy and Society.* Berlin: Gunter Roth.

Wenger, E. (1998) *Communities of Practice: Learning, Meaning and Identity.* Cambridge: Cambridge University Press.

Weston, K. (1995) Get thee to a big city: sexual imaginary and the great gay migration, *GLQ: Journal of Lesbian and Gay Studies* 2(3): 253–277.

Whatmore, S. (1999) Hybrid geographies: rethinking the human in human geography, in Massey, D., Allen, J. and Sarre, P. (eds) *Human Geography Today.* Cambridge: Polity.

Whatmore, S. (2003) *Hybrid Geographies.* London: Routledge.

Whatmore, S. and Hinchliffe, S. (2003) Living cities: making space for nature, *Soundings* 12: 12–25.

Whatmore, S. and Thorne, L. (1998) Wild(er)ness: reconfiguring the geographies of wildlife, *Transactions, Institute of British Geographers* 23(3): 435–454.

While, A. (2003) Locating art worlds: London and the making of Young British Art, *Area* 35(3): 251–263.

Wilson, A. (1972) Theoretical geography, *Transactions, Institute of British Geographers* 5(1): 37–54.

Wilson, E. (2000) *Bohemians: The Glamorous Outcasts.* New Brunswick, NJ: Rutgers University Press.

Winchester, H.P.M., Kong, L. and Dunn, K. (2003) *Landscapes: Ways of Imagining the World.* Harlow: Longman.

Wirth, L. (1928) *The Ghetto*. Chicago, IL: University of Chicago Press.

Wirth, L. (1938) Urbanism as a way of life, *American Journal of Sociology* 44(1): 1–24.

Wittfogel, C.A. (1957) *Oriental Despotism: A Comparative Study of Total Power*. New Haven, CT: Yale University Press.

Wolch, J. (2002) Anima urbis, *Progress in Human Geography* 26(6): 721–742.

Wolch, J., Emel, J. and Wilbert, C. (2003) Reanimating cultural geography, in Anderson, K., Domosh, M., Pile, S. and Thrift, N. (eds) *The Handbook of Cultural Geography*. London: Sage.

Woolley, H. and Johns, R. (2001) Skateboarding: city as playground, *Journal of Urban Design* 6(2): 211–230.

Worpole, K. (2001) *New Urban Landscapes in London: Challenges for a Post-industrial World City*. London: Groundwork.

Wylie, J. (2002) An essay on ascending Glastonbury Tor, *Geoforum* 33: 441–454.

Wyly, E. (2004) The accidental relevance of American urban geography, *Urban Geography* 25(8): 738–741.

Yakhlef, A. (2004) Global brands as embodied 'generic spaces': the example of branded chain hotels, *Space and Culture* 7: 237–248.

Yeates, M. (2001) Yesterday as tomorrow's song, *Urban Geography* 22(6): 514–529.

Young, C. and Millington, S. (2005) Living with difference, paper presented to Association of American Geographers conference, Denver, CO.

Young, L. (2003) The place of street children in Kampala's urban environment: marginalisation, resistance and acceptance in the urban environment, *Environment and Planning D: Society and Space* 21(5): 607–621.

Zukin, S. (1995) *The Cultures of Cities*. Oxford: Blackwell.

Zukin, S. (1998) Urban lifestyles: diversity and standardization in spaces of consumption, *Urban Studies* 35(5–6): 825–839.

INDEX

ability to act 146, 230
abstraction 32, 55, 77–8, 97, 103, 111, 118, 123
acquaintances 21
actants 130, 144–7, 159, 161, 205, 207, 240
Actor Network Theory (ANT) 144–7, 159, 162, 171
adaptation 95, 113, 116
advanced producer services 175, 184, 236
advertising 71, 87, 91, 101, 116; creativity 209, 217, 228–9, 231, 236
age 69
agency 38, 130, 143–5, 147, 150–1, 158–62, 216, 248
agglomeration 200–1, 228, 230, 233–4, 236
aggression 124
agoraphobia 22, 117
agriculture 28–30, 131, 133
airports 166, 168, 171, 177, 204, 217
alcohol 121, 221, 223
Algeria 154
alienation 20, 36, 61, 72, 89, 91, 117
allotments 159
Alonso, W. 29, 31
Althusser, L. 38
ambivalence 100
Amin, A. 118, 127, 204, 228, 238
Amis, M. 69

Amsterdam 184, 198, 217
anchoring points 236
Anderson, K. 151–2, 154
Anglocentrism 6
animals 152–5, 158–9, 161–2, 207, 248
anomie 20–2, 61, 63
anonymity 20, 61–2, 111
anthropocentrism 162
anthropology 25, 122, 166
anti-homeless laws 113
anti-urban myths 60, 62–3, 66–7
antisocial behaviour 17, 21, 61
Antwerp 196, 198
anxiety 22
apocalypse 129
Appadurai, A. 166
appropriation 105, 107
Arcades project 100–1
Arcadian myths 82
archaeology 131
architects 1, 31–2, 63, 77
architecture 48–9, 51, 80, 84–5; representation 88, 92
armies 131
art/artists 3, 51–2, 75, 80; creativity 211–14, 217–19, 221, 223–6, 241, 245–6
artificial intelligence 155, 157
arts 68, 80, 158, 209, 218
Ashworth, G.J. 87
Asia 11, 88–9, 181, 183–4, 242

assault 17
assemblages 147, 151
assembly lines 136
Association of American
  Geographers 4
astronomy 80
Athens 188, 211, 214
Atkinson, D. 83
Atlanta 69, 188
attitude 16, 20
Auckland 223
Augé, M. 166, 168, 177
Australia 82, 152
autobiography 122
automobiles *see* cars
avant-garde 221, 224

Babelsberg 237
Bakhtin, M. 97
balkanization 54
Baltimore 91
Bangkok 79, 184
Bank Junction 82
banking 175, 181, 241
banlieu 2–3, 61
Barcelona 87, 91
Barnes, T. 10, 46, 53
Barrell, J. 80
Barthes, R. 60
Bathelt, H. 239
Baudelaire, C. 69–70, 100–1
Baudrillard, J. 44, 72, 85
Bauman, Z. 48, 168, 170
beat poetry 3
Beaux Arts 83
Beaverstock, J. 229, 241
Beck, H. 76
begging 111, 113
Beijing 12, 189
Belgium 197
Bell, D. 215
Bengalis 242
Benjamin, W. 69, 98, 100–1, 118
Berger, J. 60, 80
Berkeley School 59, 73, 81

Berlin 19, 43, 84, 91, 216, 237
Berman, M. 35, 136, 140
Berry, B.J.L. 32
bias 73
Bible 60
bid-rent 30, 54, 207
big business *see* corporations
Big Issue 114
Bilbao 87
binge drinking 152
Binnie, J. 215, 217
biodiversity 155, 160
biology 10, 27–8, 38, 56, 118, 133, 156,
  160
biometrics 172
biotechnology 156
birds 154
Birmingham 87, 89, 91, 132, 137, 210,
  215
Birmingham School for
  Contemporary Cultural Studies 71
black boxes 143, 172, 238
Blackpool 215
Blake, M. 246
blasé attitude 16–17, 20
block-busting 42
blogs 128
Bloomsbury 225
Boardwell, J.T. 241
bobo space 225
body-subject 125
Bogard, W. 168, 170
bohemia 224–5, 230
Bondi, L. 116
boosterism 86, 92–3
Booth, C. 23
Bordeaux 195
Borden, I. 104
border controls 170, 172
BOSNYWASH 197
Boston 197, 212
boundaries 151, 154, 158, 160–1, 170
Bourdieu, P. 45, 118
bourgeoisie 12, 36, 38, 83, 224–5
Bournemouth 203

boutiques 221
Boyle, M. 89
Bradford 66–7
branch offices 236–7
branding 86–7, 168, 180, 213, 216, 233
Brasilia 78, 122, 137
Brazil 137
Breton, A. 111
Brighton 215
Brisbane 185
Bristol 159, 227
Britain 12, 22, 35, 63, 66–7
British Asians 67
Britpop 221
broadband 139, 169
brochures 79, 88
Brosseau, M. 69–70
Browne, K. 117
Brussels 196, 198
Bunnell, T. 240
bureaucracy 106, 144, 193, 245
Burgess, E. 24–6
Burgess, J.A. 73
Burgess, R. 27
Business Improvement Districts
   (BIDs) 213
business-class travel 172
butchery 153
Butler, T. 46
buzz 46, 137, 218, 232–4, 240, 245
by-laws 153
bystanders 16–17

Caen 203
calculation 18–19
California 41, 197
Callon, M. 158
Canadian Olympic Association (COA)
   188
canals 131–2, 153, 160
Canberra 82
capital cities 78
capitalism 12, 16, 19; creativity 207,
   209, 224, 226, 232; dependence
   190–2; entrepreneurs 189;

everyday 100–1, 104–5, 107–8;
   global 175–6, 179–80; hybrid 136,
   142, 164–5; theory 34–44, 46, 49,
   51–3
car parks 115, 124, 166
careers 45
Carey, J. 67
cars 124, 129, 136, 139, 143, 166,
   171
Carter, H. 58
cartography 76, 80
case studies 68, 88, 180
casinos 48
caste 14
Castells, M. 4, 35, 38–40, 58, 139–40,
   142, 176, 178–9
Castree, N. 163
cemeteries 159–60
Census 12, 29, 100
central business district (CBD) 27,
   29–30
chains of production 165
Chambers of Commerce 188
Chandigarh 78
Chapman, D. 219
Chapman, J. 219
Cherry, G. 23
Cheshire, P. 235
Chicago 6, 27, 44, 184, 213, 225
Chicago School 24–7, 50
Chicago University 24
childcare 114
Childe, G. 132
Chongqing 180
Christaller, W. 28, 32
chronological time 177
chronopolitics 138
Church, A. 203
church bells 133, 135
cinema 120, 124, 135
cinematic city 62–3
circuits 136, 234–44
circulation 136, 138, 146, 238–9
cities: definitions 1–2, 20–1, 40, 58;
   world 173–86

citizenship 244
City of Culture 89, 212
City of London 82, 202, 219, 221
city-nature formations 155
city-states 15
Ciudad Guyana 78
civil inattention 17
civilisation 65, 131, 151–2, 154, 211
Clark, D. 205
Clark, W.A.V. 54
class 13–14, 23, 26, 29; theory 35–9, 41, 43–6, 53
claustrophobia 22
cleanliness 154
Cleveland 93
clock time 18, 133, 135, 172
Cloke, P. 51, 159–60
cloning 156–7
closed circuit television (CCTV) 147, 149–51
clubs see nightclubs
clusters 228, 230, 233–8
Clyde Valley plan 122
codes 72–3, 75–6, 103, 111; hybridity 156, 158
Coe, N. 201, 240
Cohendet, P. 238
collaboration 230, 232–4
colonialism 43–4, 81, 84, 88
commerce 11, 19, 23, 30, 105
commodification 213, 216, 242
commodities 51, 85, 101, 132
communards 108
communications 2, 41, 135–8; global 169–72, 176, 179
community 15, 20, 24, 45–6, 61; creative 207, 223, 230, 237, 239–42, 244; diasporic 241–2; epistemic 175; hybrid 137
commuting 30–1, 147, 166, 198
compatibility 143
competition 24–33, 64, 93, 186; agglomeration 233; scale 190–1, 200, 202, 205

competitive advantage 93, 187, 200, 212–13, 227
competitiveness 187–8, 191, 194; scale 194, 196, 199–200, 202–3
complementarity 200, 203
complex urban networks 193
computer games 4, 61–2, 85
computers 138, 140, 142, 157–8, 166, 217, 236, 248
concentric zone model 25, 27
Concert 181
conferences 87, 91, 239
congestion 14
connections 184–5, 237, 239, 248
connectivity 165, 176, 180–1, 183, 198–9, 203, 205, 236
conservation 160
Conservative Party 65
constellations 240
construction of society 74–5, 82, 84, 98, 154
consumer-citizens 49
consumerism 56, 107, 113, 141, 154, 166, 243
consumers 36, 47–9, 54, 67; creativity 228; everyday 112
consumption 5, 34, 36, 39–40; creativity 209, 214–16, 221, 223, 245; dependence 191; everyday 101, 104–5, 122, 127; global 176; representation 70, 72, 85, 87, 92; theory 48–9, 51, 54, 58
content analysis 74
contrats de villes 196
control 49, 149–52, 173, 175, 179, 187, 245
conurbations 198
conviviality 244
Cooke, P. 41
cooking 3
cooperation 192–3, 195, 199–200, 202–3, 211, 229
Cooperation Charter 202
cooperative housing 84
Corbin, A. 133, 135

Le Corbusier 24, 61, 78, 136
corporations 51–2, 67, 83–4; creativity
    225, 236–7; everyday 112;
    representation 92; scale 175, 177,
    179, 187–8
Cosgrove, D. 80–1, 83
cosmopolitanism 46, 92, 168, 213,
    216–18, 221, 244
council housing 22, 39
council workers 33
councillors 32, 188
counterculture 105, 224
Cox, K. 189–91
craftsmen 12
creative industries 219, 229
Creative Industries Task Force
    210
creativity 3, 8, 20, 46, 64, 104, 187,
    206–46
credit rating 150
Cresswell, T. 113
Crewe, L. 218, 229
crime 2, 4, 17, 21–2, 30;
    representation 61–3, 73, 94
criminology 10
critical mass 218
Crombie, D. 188
crowds 18, 20, 22, 24, 97, 113, 121
Cruikshank, G. 219–21
cubism 3
Cultural Industries Quarter 210
Cultural Quarter 210
culture 2–4, 8, 19, 21, 145; capital 45,
    216; cars 124; creativity 208–12,
    215, 223, 228–9, 233, 235, 237, 239,
    241–6; dependence 190, 192;
    enterprise 224; everyday 95–7,
    100, 120–1, 123, 126; global 170–2,
    175–7, 179; hybrid 151–9, 161–2;
    materialism 60; metropolises 236;
    representation 59, 64–6, 68, 70–6,
    80–4, 87–8, 91, 93–4; studies 4, 6,
    10; theory 27, 46, 49, 51, 53–4, 56,
    248
curfews 113

cybercafés 140
cybertechnology 130, 155–7, 162,
    168–70, 187

Dallas/Fort Worth 47
Daly, G. 111
dance floors 121
Daniels, S. 80
Darwinism 25–6
Davidson, J. 116
Davies, G. 160–1
Davis, M. 51, 129
Davis, R. 1
De Certeau, M. 97, 104–5, 107, 121
De Vries, G. 211
Dear, M. 52–4
Deas, I.A. 188
Debord, G. 35, 97, 108, 111
debt 34
decentring 47, 49
decolonisation 43
deconstruction 60, 75–6, 93, 96,
    126
deindustrialisation 32, 86, 91, 208
Del Casino, V. 78–9
Deleuze, G. 44
Delhi 84
delinquency 14
democracy 49, 210
demography 29, 180, 214
dencentring 54
Denmark 203
deprivation 94, 98
Derby City Council 196
dérives 107, 111
Derrida, J. 44, 60, 75–6
Derudder, B. 183
desire 48–9, 64, 67, 101, 103, 124
detournement 107
Detroit 61
developers 83, 187, 191
developing world 81
developmentalism 186
Dhaka 180
dialectics 38, 70, 101, 103

diaspora 44, 241–2
Dicken, P. 195
Dickens, C. 68
difference 75, 87, 224
digital revolution 142
disabled people 81, 115–16, 127
discipline 129, 150
discourse 44, 59–60, 74–5, 103;
    analysis 74; definition 73; theory
    248
discrimination 33, 150–1
disease 2, 141
disposability 172
distance decay 30
distancing 17
distribution 34
division of labour 16, 202, 228
Döblin, A. 70
Doel, M. 123
dogma 42
domain density 236
Domosh, M. 83
Donald, J. 135
Dos Passos, J. 70
doughnut syndrome 47
downscaling 195
Drake, G. 218
drama 68, 85
dreamscape 54
Driver, F. 22
driving 95, 114, 120, 124–5
dropouts 225
drugs 49, 61, 121, 155, 217
dual-income couples 45
Dublin 87
Duncan, J. 94
Durkheim, E. 15–16, 20–1, 23
dysfunction 51
dystopias 51, 61–2

East 43
East Germany 89
East Midlands Development Agency
    (EMDA) 196
eBay 140

ecologies 24–32, 34, 39, 45; hybrid
    155, 160, 162; model 32; theory 51,
    54
economics 2–4, 8, 10–11, 18–19;
    agglomeration 228; cultural
    economies 208–9, 214, 224, 226,
    229, 234, 245; dependence 190–2;
    entrepreneurs 186; flow 202;
    global 168, 175–6, 180; rescaling
    193–4, 199–200, 203, 205; theory
    22, 28–9, 32, 37–8, 40, 45–6,
    49–50, 53–4, 56, 58
economies 3, 45–6, 53–4, 56;
    agglomeration 200–1; of
    complexity 201; creativity 209–10,
    212, 223, 230, 232, 235, 237, 246;
    dependence 190; flow 202; global
    175–7, 183, 186; hybrid 131, 164–5;
    representation 89, 91, 93; of scale
    47, 187, 200–1; of scope 47, 187,
    200; twenty-four-hour 173;
    weightless 137, 210
economists 1, 23, 201
Edensor, T. 116
edge cities 47, 54, 166
Edinburgh Festival 87
education 4, 6, 39, 88; creativity 210,
    212, 215, 217, 233; representation
    93
electricity 132–3, 141–3, 189
elites 4, 11, 23, 92, 131–3; global 176–7
Ellis, B.E. 69
emancipation 20
embodiment 97, 116
emerging markets 179
Emilia-Romagna 228
Emin, T. 219
empathy 21
employment 45, 93, 208
encounter 118
Engels, F. 16, 37–8, 50
engineering 136, 142
England 12, 66, 89, 195–6, 203, 219,
    227
English language 11

entertainment 46
entrepreneurs 51, 65, 86, 132; creativity 212, 217–18, 225, 241
entropy 31
environment 22, 24, 54, 59; representation 80, 87
ergonomics 10
estate agents 32–3
ethics 161, 246
ethnicity 4, 13, 26, 29, 33; creativity 213–14, 240–2, 244; representation 76, 81; theory 43–4, 49, 54
ethnography 25–7, 68, 121–2, 144
etiquette 17
eugenics 25
Eurocentrism 205, 209
Eurocities 196
Europe 12, 47, 66, 87–9, 160; creativity 212, 217, 235; hybrid 162; scale 173, 181, 183–4, 195–8, 202
European Commission 196
European Union 203
events 113–21
everyday city 95–128
everyday creativity 218–26
evictions 112
evolution 26, 42, 54, 142, 151, 171
exchange 34, 156, 199
exchange value 18, 36
exclusion 49, 73, 81, 91
expatriates 241
expressive specialists 241
external city 91
eye contact 116

factorial analysis 29
factories 132–3, 212
fashion 3, 18, 64, 84, 105; creativity 209, 219, 236, 241
fast-tracking 172
fear 2, 243
feminism 4, 42–4, 52–3, 81, 94, 114, 117, 157
feudalism 12, 16, 34, 66, 103

Fielding, H. 69
film noir 62–3, 129
film-making 63, 209, 228, 237, 241
Financial Deregulation Act 202
financial institutions 32, 51, 92
financial services 236, 241
firm formation 201
first-class travel 172
fixities 190–1, 242
flagship projects 91
flâneurs 101–2, 128, 218
Flemish Diamond 196–8
flightpaths 234–44
Florence 214
Florida, R. 187, 212–15, 218, 233, 246
flow 136–41, 153, 165–6; air travel 183–4; creativity 221, 235; dependence 190–2; economic 202; global 179–81, 185; information 201; rescaling 198, 203, 205; scale 168–70, 175–8, 180, 204
Flowerdew, J. 88
Flusty, S. 54, 105–6
food 131, 154, 223, 241
Fordism 47, 136
formations 159
Foucault, M. 44–5, 60, 74, 76, 106
foundational story 151
fourth world 177
fox hunting 66
France 23, 108, 133, 135
Frankfurt 173, 179, 181, 183–4
free-running 107
freedom 20, 43, 48–9
freeways 136
French revolution 66
friction of distance 169
Friedmann, J. 173
fronts 18
fuzzy set theory 31
Fyfe, N. 122, 127

galleries 219–20
Gandy, M. 141, 157

gangs 26, 61
Garden City 61
Garreau, J. 47
gas 132–3, 141, 143, 189
gated communities 47, 147
Gaudí, A. 87
gay identities 45, 68, 112, 117, 214–17, 242
gaze 45, 63, 102, 106, 224
Geddes, P. 173
gemeinschaft 15, 21
gender 4, 14, 43, 52, 69, 81; creativity 246; everyday 102, 114, 116–17; hybrid 157
generative aspects 2
genetic engineering 66, 156
genius loci 69
gentrification 4, 42, 45–6, 56; creativity 214, 221, 225–6, 236; everyday 112; representation 92, 94; scale 187
geographers 1, 4, 7, 9–11, 26
geography 4–8, 10–11, 13–14; theory 27–9, 33, 36, 43–4, 46, 56–8, 248
geopolitics 124, 138, 154
George, H. 23
Gere, Richard 65
Germany 197, 203
gesellschaft 15
Ghent 196
ghettos 25, 29
Giddens, A. 166
Giordano, B. 188
Glaeser, E. 228
Glasgow 89, 91, 93, 122, 132, 137, 198, 212
global chains 168
global cities 180, 185
global service firms 174, 180–3
globalisation 32, 44, 165, 170; scale 172–86, 190, 192, 202, 204–5
Globalisation and World City (GaWC) 180–1, 183
god's eye view 97

Goffman, E. 18
Gold, J. 73
golden ages 211
Goldsmiths College 219, 226
Gornostaeva, G. 235
Goss, J. 84
gossip 223, 228, 239
Gottmann, Jean 197
government 125, 144
Grabher, G. 229, 238
grafitti 111
Graham, B. 91
Graham, S. 141–2, 150, 162
Gramsci, A. 70–1
grass 153, 159
gravity models 28
Greater London Authority (GLA) 193–4
Greater Manchester Council (GMC) 195
Greed, C. 114
Greeks 20, 64
Grenfell, M. 226
Grenoble 196, 212
Griffiths, H. 155
Gross Domestic Product (GDP) 210
Grosz, E. 117
Groundwork UK 160
group consciousness 233
growth coalitions 188–90, 193
Guggenheim Museum 87
guidebooks 79
guilds 12
guns 61
gypsies 154

habitation 122
Hague 198
Hall, P. 173, 197, 211–12, 214, 246
Hall, S. 73
Hall, T. 58
Hamilton 117
Hanna, M. 78–9
Hannam, K. 94
Hannerz, U. 241

Hanson, S. 246
Haraway, D. 156
Harding, A. 192
Hardy, C. 226
Harley, B. 76
Harris, C. 27
Harvey, D. 28, 33, 35, 38
Haussmann, G. 69, 78, 136
Le Havre 203
Haythornthwaite, C. 142
health 22, 39
hegemony 43, 71, 94
Heidgegger, M. 118
Hemel Hempstead 201
heritage parks 48, 85, 91
Hermant, E. 97, 141, 146–7
heterotopias 242
hierarchies of hybrid 164–5
Highmore, B. 100
Hinchliffe, S. 151, 155, 160
hinterland 203
hinterlands 183
Hirst, D 219, 226
historians 80, 133
historical materialism 55, 101, 103
history 10, 151–2, 179, 196, 242
hobos 26
Hochschild, A.R. 223
Hoggart, R. 70
Hollywood 63, 65, 236
Holston, J. 122, 137
Home, Ian 109
homeless 49–50, 111–13, 127, 147, 187
homeowners 30
homophobia 33
Hong Kong 88–9, 184
Hornby, N. 69
hospitality 210
hospitals 39, 144
hotels 168, 177
house prices 112
housework 114
housing 4, 19, 22, 26–7; theory 30–4,
    38–9, 41, 45–6
Howard, E. 61

Hoxton 219–21, 225–6
Hoyt, Homer 27
HSBC 181
Hughes, C.G. 89
humanities 10, 68
Hume, G. 226
husbandry 80
hybrid city 129–63, 244
hypermobility 165–73, 204, 212
hyperreality 72, 85
hypotheses 10
hysteria 22

iconography 72, 79–81, 84, 91, 93,
    152
ideology 13, 38–9, 44, 51; theory 55–6
imagination 60–8, 93, 121, 129–30
immanence 127, 150
imperialism 43–4, 81, 83–4
impressionism 3
improvement 112
improvisation 95–6, 107, 118
indexical 72
indifference 16–17, 21
individualism 15, 19–20, 49, 67, 230
industrial revolution 12, 132
industrialisation 15–16, 19
industry 3, 6, 29, 46–7, 91, 228
inequality 23, 33–5, 43, 51, 73, 89,
    93–4, 111
inertia society 139
information 137, 178, 209
information and communications
    technologies (ICT) 168–70, 232–3,
    236
infrastructure 41, 131–43; global
    176–7
inhuman metropolis 20–4
injustice 96
inner cities 45–6, 73, 213, 215
innovation 3, 20, 41, 129–31; creativity
    209–10, 212, 221, 223, 226,
    228–30, 232; entrepreneurs 186;
    types 234–6, 239–40, 246
insects 155

Institute of British Geographers 4
integration 2
intellectual property 210
intentionality 119, 160
interdependence 144, 198, 201, 228
interlockers 181
internal city 91
Internet 75, 137–8, 140, 142–3, 162, 169–70, 212, 236
Interreg IIIB 203
interruptions 111
interventions 23, 32–42, 107, 127, 234
intransitive city 164–205
introversion 22
investment 40–1, 45, 85–9, 92
Ireland 85
Irigaray, L. 44, 117
Irish pubs 85
irrigation 131–2
Islam 66
Italy 83, 197, 228
itinerant traders 49

Jackson, R. 94
Jacobs, J.M. 82
Jameson, F. 52
Jansson, J. 230–1, 233
Japan 175, 197
Jarvis, H. 114
jazz 3
Jessop, B. 186
Jews 214, 241
job creation 47, 210
job insecurity 48
Johnston, R.J. 4
Jones, J.P. 58–9
Jones, O. 159–60
Jopling, J. 226
Joyce, J. 70, 87
just-in-time production 173

Kaika, M. 130, 153
Kanasai 197
Karez wells 131
Kavaratzis, M. 87

Keeble, D. 235
keno-capitalism 54
Khartoum 180
Kim, Y.H. 173, 205
King, A. 84
kinship 15, 18, 20–1
knowledge 3, 7–8, 43, 45; creativity 210, 212, 223, 228–30, 232–5, 237–9, 245; everyday 95, 98, 110, 119; global 181, 187; hybrid 137, 143–4, 152, 159
Knox, P. 58, 83, 94, 174–5
Koskela, H. 147, 150
Kostof, S. 58
Kracauer, S. 19
Kraft-Ebing, R. 23
Krätke, S. 216, 237
Kresl, P.K. 199–200
Krugman, P. 200, 233
Krutnik, F. 63
Kureishi, H. 66, 69

laboratories 24–5
labour 35–8, 40–2, 91, 103
Labour Party 193–4, 210, 244
L'Aire Urbaine de Lyon 195
laissez-faire 195
Lake, R.W. 55
Lambooy, J.G. 198
Lanchester, J. 69
land prices 41
land use 24–32, 41
Landry, C. 246
landscape 39–41, 43, 51, 54; representation 59, 61, 67–8, 72, 79–84, 87, 91–4
Landy, M. 224, 226
Lane, A. 226
language 10–11, 27–9, 44, 53
laptops 139
Latham, A. 122–5, 221, 223, 225
Latour, B. 97, 141, 143–7
Laurier, E. 121
Law, J. 144
laws 6, 9, 28, 88, 105, 107, 111, 113, 175

leadership 108, 175, 189
league tables 86
learning effects 228
Lees, L. 121–3
Lefebvre, H. 35, 77, 97, 101
Left Bank 225
LeGates, R.T. 58
Leicester 89, 196
Leicestershire Promotions 196
leisure 5, 39, 47, 49, 72; creativity 241;
    everyday 102, 116, 127
Leitner, H. 92–3
lesbian identities 45, 117, 214–16, 242
Lessing, D. 69
lethal response 150
Leuven 196
Ley, D. 45, 84, 94
libido 23
life-support systems 157
lifestyle 46, 60, 91, 166, 215, 219,
    224
lighting 115, 129, 132–3, 135, 141, 152,
    157, 162
Lille 198
liquidity 170
Lisbon 91
literature 10, 69–70, 73
Liverpool 89
livestock 153
Living Spaces 160
Lloyd, R. 213
local area networks (LANs) 139
local authorities 193, 203
local dependence 189–90, 192
localities 164–5
locational theory 28
locomotion problems 116
Lofland, L. 17
Logan, J.R. 188
London 12, 23, 43, 46, 49; creativity
    211, 214–15, 219–21, 224–6,
    228–9, 235–7, 242, 246; Plan 193;
    representation 65, 68–9, 76–8, 82,
    84, 88, 91; scale 173, 175, 177,
    179–81, 183–5, 195, 197, 201–2,
204; Underground map 76–7; as
    world city 208–9
London and Westminster Chartered
    Gas Light Company 132
Longhurst, R. 117, 127
Lorimer, H. 120, 126
Los Angeles 6, 49–51, 54, 57
Lösch, - 28
Lucas, S. 219, 226
Lynch, K. 31
Lyon 195–6, 202–3
Lyon, D. 150

McCann, E. 104, 185
McCormack, D. 122–5
McEwan, I. 69
McInerney, J. 69
MacKay, - 26
Mackintosh, C.R. 89
MacLeod, G. 192
McLuhan, M. 170
McNeill, D. 92
Macs 143
Madrid 173
malls 47–8, 67, 72, 84–5
Malmberg, A. 233
managerialism 32–4, 42
Manchester 3, 37, 132, 185, 195–6,
    212, 215–18
manners 152
Manningham 67
mapping 28–9, 69, 76–9, 81;
    representation 93
marginalisation 115
market economy 26, 30
marketing 60, 86–7, 89, 91–3;
    creativity 212–13, 216–17, 233,
    242
Marketing Manchester 217
marketplaces 48, 72
Markusen, A. 214
Marsden, T. 145
Marseille 195, 202–3
Marvin, S. 141–2, 162
Marx, K. 13, 20, 35–8, 40

Marxism 32–46, 49, 53–5; everyday 97; hybrid 145; representation 71–3, 80, 85; scale 178
Maskell, P. 233
Massey, D. 52, 179
materiality 96, 121–6, 130
mathematics 10, 28, 38, 56, 80
Matless, 122
The Matrix 67
Mauss, - 118
May, T. 4
mechanical solidarity 15–16
media 4, 59–63, 65, 68; creativity 209, 213, 216, 221, 225, 235–7, 246; representation 70–1, 73, 88, 93
Media Quarter 210
mediators 121
mega-cities 180
megalopolis 197
mental illness 22
mental maps 31
merchants 12
Merleau-Ponty, M. 118–19
Merrifield, A. 58
Merriman, P. 125
Mesopotamia 11, 131
metallurgy 132
metanarrative 55
Metics 214
Miami 179, 184
miasmas 22–3
Michelangelo 214
migration 63, 65, 168, 190, 204, 214, 240–1
Mikesell, M. 81
Milan 174, 184, 195, 219
Miles, M. 100
Milgram, S. 17
Milicevic, A.S. 34
military 11, 132
Milton Keynes 201
mimesis 70, 86
mind-body link 119
Minneapolis 198
Mitchell, W. 139, 141, 162

mixity 243–4
mobility 2, 12, 43, 114; scale 165–73, 189, 191, 204
modelling 27–9, 32, 54–5, 100, 121
modernism 6, 8–58, 78, 135–7, 206–7, 245, 248
modes of production 34, 36, 38
Molotch, H. 188, 216, 218
Monaco Grand Prix 87
money 18–19, 23, 36, 67
Moorcock, M. 69
Moore, R. 33
morality 20, 23, 25–6, 61
Morel, P. 61
Morgan Stanley Dean Witter 181
morphology 2, 176–7, 203
mortgage lenders 32
Moses, R. 136–7
motorways 125, 137
movement 118
Mozart, W.A. 214
multi-tiering 193
multiculturalism 87, 91, 224, 244
multimedia 236
multinationals 51–2, 237
multiple nuclei model 27
multiplexes 48, 62
Murdoch, J. 145, 158
museums 48, 226
music 3, 5, 61, 64, 75; creativity 209, 213, 217, 221, 230–4, 241
Musterd, S. 198
My Son the Fanatic 66–7
mystification 52
myths 60–8, 91, 212, 237, 242

Nachum, L. 235
narrative 63, 67, 73, 86, 93, 122, 135, 186
nationality 13, 81
Natter, W. 59
natural sciences 10
nature 2, 8, 19, 23–4, 133, 135, 151–61, 248

navigation 31, 114
Nead, L. 78
needs 48
Neff, G. 230
neo-Marshallian nodes 228, 246
neo-urbanism 61
Netherlands 197, 203
networks 142–7, 153–4, 159; complex
    202–3; creativity 207–8, 221, 223,
    225–37, 239–40, 244–5; global
    175–7, 179–81, 183–4; hybrid
    161–2; theory 248
Neural Network Face Recognition
    149
neurasthenia 22
neuroses 22
new economic geography 201
new mobilities paradigm 170–2
new order 136, 142
New Orleans 3
New York 12, 69, 83, 88; as world city
    208–9
New Zealand 117, 223–4
Newton, Isaac 28
Nicholson, G. 69
nightclubs 48, 217, 221, 223, 228
nightlife 79, 88, 121, 133
no-go areas 4
noise 22, 116
non-capitalist industries 209
non-consumers 49
non-places 139, 168, 177
non-representational theory 119–21,
    126–7, 171
North Africa 154
North America 6, 81, 155, 162, 173,
    181, 184, 212
North Sea 203
North Western Europe 198
Northern Ireland Assembly 194
Norway 203
Nottingham 89, 196
Nottingham City Council 196
Nottingham East Midlands Airport
    196

novels 68–70, 74
nuclear households 15
Nye, D. 133

obesity 118
observation decks 98
Office of the Deputy Prime Minister
    196
offices 47, 139–40
Ofili, C. 219
oil 124
Olympic Games 188–9
ontology 130, 145, 162, 245–6, 248
Open University 58, 205
Ordnance Survey maps 78
organic solidarity 16
Orient 43
Osaka 174
Other 42–55, 63, 112–13, 154, 216–17
outsourcing 47

pace of life 135–6, 140, 173
Pacific Asia 181, 183–4
Pacific Rim 175
Padua 197
Pahl, R.E. 33
panhandling 113
panoptics 98, 106
Parent-Duchatelet, A. 23
Paris 2–3, 33, 41, 43; arcades 100–1;
    as world city 208–9
Park, R.E. 24–6
Park, S. 226
parking lots 104, 113
parkouristes 107, 111
parks 104, 112–13, 118, 147, 153
Parr, J. 199–201
Patch, J. 214, 226
pathologies 20–4, 26
patriarchy 43
patrons 226
pedagogy 27, 164
pedestrians 98, 106–7, 114, 122
Peirce, C.S. 71
performance 97, 144

periphery 47, 61
Perl, J. 246
Perry, B. 4
Peterborough 201
Phelps, N.A. 200–1
phenomenology 118–20, 125
Philo, C. 152
phobias 22
photography 75, 80, 135, 241
physics 10, 28, 139
physiognomy 150
Pike, D. 76, 78
Pile, S. 2
Pinch, S. 58
Pinder, D. 109
Pittsburgh 93
planners 1, 31–2, 41–2, 63
planning 6–7, 23–4, 29, 34; everyday
    100, 114
plants 153–4, 159, 248
Plato 214
play 92, 105–8, 138, 219
pluralism 88, 124–5
Pocock, D. 69
poetry 68–9, 71, 100–3, 122,
    128
polarisation 54, 92
police 32, 49, 106, 112–13, 150–1,
    187–8, 190
policy 4, 29, 34, 86
polis 138
political economy 34, 49; theory 56,
    248
politicians 78, 86, 144, 146, 151, 195,
    216, 226
politics 3–4, 7–8, 10–11; cooperation
    203; global 170, 175, 186; theory
    33, 35, 38, 42, 44, 46, 50–1, 53–4,
    56, 58
pollution 14, 22, 89, 112, 124, 153–4,
    245
polycentric linking 193
polycentric urban regions (PUR)
    196–9, 203
polymorphous 130

polysemous 66, 71
Poole 203
Poovey, M. 78
population density 1–2, 12, 21, 131–2,
    208, 211
Porter, M. 199, 233
Portsmouth 203
positivism 28, 30, 33–4, 42, 55
post-Fordism 47, 69, 84–5
post-human cities 126, 129, 151–62,
    234
post-industrialisation 47–9, 61, 86,
    89, 91, 132, 210, 230
post-production 235–6
post-structuralism 42–55, 74–5, 85,
    94, 123, 146, 205
postcards 79, 85–6
postcolonialism 43–4, 81, 84, 204
postmodernism 6, 8–58, 84, 96, 177,
    206–7, 245, 248
Poundbury, Devon 61
poverty 2, 14, 23, 33, 37
power 13, 34, 43–4, 66
Power, D. 230–1, 233
practice 95–7, 100, 103, 107; creativity
    207–9, 218, 221, 225–6, 239–40;
    everyday 113–14, 116, 119–22, 124,
    126–7; hybrid 161; theory 247
Prasad, U. 66
Pratt, G. 67
pregnancy 117
Prestige Projects 87
privacy 17
private sector 88, 92, 186–9
privatism 243
privatopias 47, 54, 147
production 34, 36–9, 44; creativity
    209, 211, 213, 215, 218–19, 221,
    226–8, 230, 234–6, 238, 245;
    dependence 191; entrepreneurs
    186–92; everyday 101, 105, 122;
    global 172–3, 176; theory 47–9, 56
productivity 18, 115
profit 36–7, 40–1, 47, 51, 140, 175,
    186, 226

project work 229–30
proletariat 36–7
promotion 86–9, 91–3, 112, 186–7, 210
property 4, 40, 112, 192; consortia 188; developers 51; markets 32–3; prices 64, 225
prosthetics 156
prostitutes 23, 25–6, 49, 65, 67, 101, 113
Protestantism 224
psychiatrists 22
psychogeographies 108–9, 128
psychology 10, 16, 22
public housing 3
public sector 88, 92, 188–9
public-private partnerships 92, 188
pubs 85, 221, 223
Pulitzer, J. 83–4
Puri, O. 66
Puritans 133

queer theory 44

race 14, 43, 69, 81, 214, 218
racism 3, 29, 33, 73, 150, 154, 245
Rae, F. 226
railways 132, 138, 143, 153, 160, 162, 166, 217
ramps 116
Randels, S. 195
Randstad 197–8
rap music 61
rape 17
rapid-transport systems 198
reading cities 60, 68–79, 93, 98, 247
real estate 27, 85, 111, 191, 226
recession 37
Reckless, W.C. 25–6
Reclaim the Streets 107
redevelopment 85, 100–1, 115, 137
Reeves, K. 67
referents 72, 85
reflexivity 175

reform 23, 153
refugees 241
Région Urbaine de Lyon (RUL) 195
Regional Chambers 194
Regional Development Agencies 194
regionalism 194
Reid, P. 203
reification 209
relationality 204, 238
religion 13, 67, 76, 132, 244
Relph, E. 177
Renaissance 80, 83
rent gap 41, 45
representation 44, 53, 57, 72
represented city 59–94
repression 49, 68, 100, 104
rescaling 192–203
rescripting 221
Le Reseau de Villes 196
resistance 51, 100, 106–8, 121
rest 118
restaurants 221
revanchism 112–13
revitalisation 96
revolution 35, 37–8, 66; everyday 100–1, 103, 108, 126
rewriting cities 86–93
Rex, J.A. 33
rhetoric 187, 194
Rhine-Ruhr 197–8
riots 2–3, 33, 66, 73, 107–8
rituals 152, 154, 173
rivalry 233
roads 124–5, 138, 142, 153, 159, 166, 203
Roberts, J. 65
Robins, K. 136–7
Robinson, J. 185, 205
robotics 155, 158
Romans 20, 64
Rome 83, 184
Rose, G. 94, 100
Ross, K. 154
Rotterdam 198
Roubaix 198

Rouen 203
Rowson, M. 221
Royal Zoological Society of South Australia 152
rural areas 1–2, 7, 12, 14–18; theory 20–1, 57
Ruskin, J. 60
Russia 35
Russian revolution 66
rust belt 86

Saatchi, C. 226
Said, E. 43
St-Etienne 195, 196
St Paul 198
San Diego 212
San Francisco 3, 66, 174, 184, 225
San Juan, R.M. 67
Sanitation Department 112
SARS virus 89
Sassen, S. 173, 175–6, 180, 184, 202, 205
satellite technology 97–8, 124, 129, 170
Sauer, C. 81
Saunders, P. 1
Saussure, F. 71
Savage, M. 14, 53, 58
scale 164–5, 189; intransitive 191–3, 200, 204
scanscape 149
scene 223, 232, 234
Schivelbusch, W. 135, 162
schizophrenia 22, 178
schools 39, 114
Schuldt, N. 239
science 28–9, 130–1, 143–5; parks 47, 229; social production of 144–7
science fiction 62–3, 155–6
science, technology and society (STS) 143, 238
Scott, A. 41, 223, 228, 246
Scott, J. 78
Scottish Parliament 194
scripts 217

Seamon, D. 118–19
Searle, A. 219
Seaside, Florida 61, 67
seasons 19
Seattle 3, 225
Second World War 115
sector model 27
sedition 133
selective porosity 153
semiotics 71–2, 75, 94
Sennet, R. 128
sensory bombardment 22
Seoul 173
Serres, M. 145–6
service sector 201, 210, 241
sewage 23, 132, 141, 143, 153, 166
sexism 33, 43, 117
sexologists 23
sexuality 4–5, 27, 65, 67–9
Shakespeare, W. 214
Shaw, 26
Sheffield 210
Sheppard, E. 92–3
shop window culture 19
shopping 19, 100, 104, 106; everyday 113–15, 117
Short, J. 58, 62, 66, 87
Shurmer-Smith, P. 94
signage 116
signifiers/signified 71–2
Silicon Valley 227, 240
Simmel, G. 16, 18–19, 23
Sinclair, Iain 109, 128
Singapore 173, 175, 179
situationism 107–9, 111
Situationist International (SI) 108
skateboarding 104, 106–7
Skeggs, B. 217
skyscrapers 83, 88
slaughterhouses 153
sleeping 112–13
slogans 181
Slow Cities 160
slump 40
smallholdings 159

Smith, N. 41–2, 45, 112, 187
Smith, R. 205, 238
Smith, Z. 69
Smithfield market 153
social action 97, 118
social relations 8, 14; everyday 122, 144–5; theory 21–2, 36, 38, 40
social reproduction 114, 191, 205
social sciences 4, 10
social services 39
socialism 34–5
sociality 223, 225, 234
sociologists 1, 15–16, 20, 22–3, 57, 61
sociology 4, 6–7, 10–11, 14, 24–6, 32–4, 178
Sodom 60–1
software 3, 143, 150–1, 159, 161, 209, 231, 248
Sohonet 236
Soja, E. 2, 49–50, 52–3
Solnit, R. 225–6
Southampton 203
space 40, 44, 49, 57, 59–60; agglomeration 233; bobo 225; branded 168; conceptual 76, 78; creative 207, 238, 244; definition 238; devalued 178; diasporic 241–2; differential 224; diseconomies 201; domination 104; economy 202; embodied 113–21; entrepreneurial 189–90; everyday 104; fix 38; fluid 165; forms 103; geometry 78; global 172, 176–7, 180; hybrid 139–40, 153; interdictory 243; meaning 73; politics 138; public 111–12; radical 108; relations 57, 83; representation 63, 67–8, 70, 72, 105; resistance 106–7; science 24–32; subversion 111; surveillance 147; tourist 79
space-time: compression/ distanciation 166, 168, 177

Spatial Development Strategy for London 193
speed 138–9, 141, 148, 166, 168, 171, 173
sport 87, 166, 168, 210, 217
stage sets 63
stairs 116
Stalinism 35
state 39–42, 78, 82, 92; scale 164, 187, 191, 194
statistics 23, 28–30, 175, 180, 215
stereotypes 65, 73, 215
stigmatisation 3, 112
stock exchanges 142, 175
Stockholm 230–4
Stoke 210
Storper, M. 56, 175, 228, 232–3, 246
Stout, F. 58
strangers 17–18
street(s) 101, 104, 122, 224; ballet 114; children 111, 127; control 187; furniture 146–7; interactions 18; level 106; life 242–3; lighting 132–3, 135, 152; parties 107; surveillance 147
structure 39, 42, 44, 46; theory 53, 55, 248
students 66, 88, 108, 145, 213, 221, 225
subaltern theory 44, 53
subcontracting 47
subcultures 25–6
subsidiaries 236–7
suburbs 41, 213
Sumeria 11
Sun Belt 197
superheroes 62
supermodernity 168
surface culture 19
surgery 78
surplus 36, 131
surrealism 111
surveillance 17–18, 49, 67; everyday 97, 114; networks of 135, 138, 147, 150–1

surveying 78, 80
Susser, I. 58
Sweden 78, 203, 231–2
Swyngedouw, E. 130, 154
Sydney 173, 185
symbiosis 26, 223
symbols 59–60, 79–80, 82
synergy 201–2
syntax 10

tacit knowledge 229–30, 234, 238–9
talking 95, 118, 121
task-orientated work 133
taxation 189
Taylor, P. 164, 179–81, 185, 205
technocrats 132
technology 3, 5, 12; creativity 211–12,
    215, 228, 231–2, 245; determinism
    141–2; everyday 124; flow 130;
    theory 28, 36, 41, 248
technopole 196
technopolises 175
technoscience 156
telephone queuing 150
tempo 136
terrorism 51
text 60, 68, 70–1, 73; representation
    75, 79–81, 83–4, 86, 89, 93–4
textiles 3, 132
textures 95
Thailand 79
theme parks 85
theory, definition 10–11
Theran 180
Third Italy 47
Third World 241
Thomas, W.I. 25
threats 18, 46, 63, 153
Three Cities Town Net 196
Thrift, N. 4, 56, 100, 116
thuggery 14
time 18–19, 37, 40, 61; instantaneous
    172
time-orientated work 133
Tivers, J. 114

toilet facilities 116–17
Tokyo 173, 175, 177, 179–81, 202,
    228
Tonnies, F. 15, 20
topography 54, 68, 76, 107, 152
topology 145, 170–1, 176, 238
Toronto 188–9
Tourcoing 198
tourism 78–9, 87–8, 91, 98, 112, 210,
    220, 241–2
Town Net 203
townscape 83
Townsend, A.R. 201
trade 11, 65, 88–9, 168–9, 173, 214
trade fairs 239
trade unions 189
traffic 124, 136–7
transgression 111–13
translation 179
Transmanche Metropole 203
transnational corporations (TNCs)
    175, 179, 195, 236
transnational lifestyles 2
transnationalism 192, 240–1
transport 2, 12, 27, 41
transsexual identities 215
transvestism 117
travel 5, 22, 31, 114; air 183–4;
    weightless 168
travel-to-work 190, 192, 196
trees 153, 159–60
Treviso 197
trialectics 103
The Truman Show 67
Tufts, S. 188
Turk, G. 219
twenty-four-hour society 173
twilight zone 26
twinning 203

UBS Walburg 181
Uganda 111
UK Core Cities 196
Ullman, E. 27
umbrella tier 193

underclass 23, 41
underpasses 114–15, 124
unemployment 48, 65, 93, 112
United Kingdom (UK) 12, 65
United States (US) 24, 29, 47
universities 145, 147, 188
upscaling 192, 202
ur-city 50
urban decay 60, 89
urban land nexus 41
urban regimes 188–91
urban renewal 38, 40, 64, 91
urban studies 7, 24, 34, 40; theory 42,
    45, 248
urban theory 1, 3–4, 6–58
urbanisation 11–19, 44, 108, 132, 173,
    185
urbanism 1, 42, 45–6, 51
urbanity 224
Urry, J. 124, 170–3
use value 18–19, 36, 104
utility theory 31

vagrancy 14
values 16
Van Zelm, I. 198
Vancouver 84, 175
Venables, A.J. 228, 232
Venice 197
Venice Biennale 87
vice 62
Victorians 23, 60, 68, 160
Vienna 84, 211, 214
viewing platforms 98
violence 49, 61, 96
Virilio, P. 129, 137–9
virtualism 137–9, 234
visual impairment 116
visualities 100
vocabularies 10–11, 26, 247
Von Thünen, J. 28–9

Wagner, P. 81
Wales 12, 194
Wales, Prince of 82

walking 95, 98, 100, 107, 113, 120–1
Wallerstein, I. 179
wants 48
warehouses 212, 219
Washington 184, 197, 212
water 131–2, 141, 143, 147, 153–4, 157,
    189, 211
waterfronts 91
wayfinding 31
Wearing, G. 219
webcams 172
Weber, M. 20, 23, 28, 32, 211
websites 75, 86, 128, 145
Webster, F. 136
weeds 153, 159
weight 118
Weimar Republic 19
welfare 41, 116, 187
Wellman, B. 142
Welsh Assembly 194
West 4, 6, 33, 43–4, 47
Westminster City Council 236
Whatmore, S. 146, 160, 163
wheelchairs 116
While, A. 221, 224
White Cube Gallery 219–20, 226–7
white flight 29
Whiteread, R. 219
Williams, R. 60, 70, 80
Winchester, H.P.M. 94
Windows 143
wired city 169
wireless zones 236
Wirth, L. 21–3, 25, 208
withdrawal 22
Wittfogel, C.A. 131–2, 141
Wittgenstein, L. 118
Wolch, J. 152, 155, 158
Wolfe, T. 69
women 20, 43, 63, 107, 114–18
word-of-mouth 235
work ethic 133, 224
World Building 83–4
world cities 88, 173–86, 202, 205,
    208–9, 228, 235–7

world systems theory 179
World Trade Center 98
World Wide Web 142, 169
worlding 245
Wright, F.L. 24
writing cities 80–93, 98, 102, 129, 247
Wyly, E. 57

Yakhlef, A. 168

Young British Artists (YBAs) 219, 224–6
Young, L. 111
youths 49

zero tolerance 49, 113, 187
zoos 152
Zukin, S. 84, 221

Lightning Source UK Ltd.
Milton Keynes UK
UKOW05f0911101013

218789UK00001B/13/P